Sixten Boeck

Development and Application of the S/PHI/nX Library

Sixten Boeck

Development and Application of the S/PHI/nX Library

First-principles Calculations of Thermodynamic Properties of III-V Semiconductors

Südwestdeutscher Verlag für Hochschulschriften

Impressum/Imprint (nur für Deutschland/ only for Germany)
Bibliografische Information der Deutschen Nationalbibliothek: Die Deutsche Nationalbibliothek verzeichnet diese Publikation in der Deutschen Nationalbibliografie; detaillierte bibliografische Daten sind im Internet über http://dnb.d-nb.de abrufbar.
Alle in diesem Buch genannten Marken und Produktnamen unterliegen warenzeichen-, marken- oder patentrechtlichem Schutz bzw. sind Warenzeichen oder eingetragene Warenzeichen der jeweiligen Inhaber. Die Wiedergabe von Marken, Produktnamen, Gebrauchsnamen, Handelsnamen, Warenbezeichnungen u.s.w. in diesem Werk berechtigt auch ohne besondere Kennzeichnung nicht zu der Annahme, dass solche Namen im Sinne der Warenzeichen- und Markenschutzgesetzgebung als frei zu betrachten wären und daher von jedermann benutzt werden dürften.

Verlag: Südwestdeutscher Verlag für Hochschulschriften Aktiengesellschaft & Co. KG
Dudweiler Landstr. 99, 66123 Saarbrücken, Deutschland
Telefon +49 681 37 20 271-1, Telefax +49 681 37 20 271-0
Email: info@svh-verlag.de
Zugl.: Universität Paderborn, Diss., 2009

Herstellung in Deutschland:
Schaltungsdienst Lange o.H.G., Berlin
Books on Demand GmbH, Norderstedt
Reha GmbH, Saarbrücken
Amazon Distribution GmbH, Leipzig
ISBN: 978-3-8381-1276-3

Imprint (only for USA, GB)
Bibliographic information published by the Deutsche Nationalbibliothek: The Deutsche Nationalbibliothek lists this publication in the Deutsche Nationalbibliografie; detailed bibliographic data are available in the Internet at http://dnb.d-nb.de.
Any brand names and product names mentioned in this book are subject to trademark, brand or patent protection and are trademarks or registered trademarks of their respective holders. The use of brand names, product names, common names, trade names, product descriptions etc. even without a particular marking in this works is in no way to be construed to mean that such names may be regarded as unrestricted in respect of trademark and brand protection legislation and could thus be used by anyone.

Publisher: Südwestdeutscher Verlag für Hochschulschriften Aktiengesellschaft & Co. KG
Dudweiler Landstr. 99, 66123 Saarbrücken, Germany
Phone +49 681 37 20 271-1, Fax +49 681 37 20 271-0
Email: info@svh-verlag.de

Printed in the U.S.A.
Printed in the U.K. by (see last page)
ISBN: 978-3-8381-1276-3

Copyright © 2010 by the author and Südwestdeutscher Verlag für Hochschulschriften Aktiengesellschaft & Co. KG and licensors
All rights reserved. Saarbrücken 2010

Contents

1 Theory 9
- 1.1 The Many-body problem 9
  - 1.1.1 Born-Oppenheimer approximation 10
  - 1.1.2 Electron-electron interaction 11
- 1.2 Density functional theory 12
  - 1.2.1 Kohn-Sham formalism 12
  - 1.2.2 Kohn-Sham equations 13
  - 1.2.3 XC functional 16
- 1.3 Periodic boundary conditions 17
- 1.4 Integration over the Brillouin zone 18
- 1.5 Valence/core partitioning 23
  - 1.5.1 Pseudo-potential theory 24
  - 1.5.2 All-electron approaches 27
- 1.6 Tight-binding methods 31
- 1.7 Forces in ionic systems 32
  - 1.7.1 Hellmann-Feynman theorem 33
- 1.8 Conclusions 34

2 Methods 35
- 2.1 Electronic minimization schemes 35
  - 2.1.1 Gradients $-\frac{\delta}{\delta\langle\Psi|}$ 36
  - 2.1.2 Search direction 38
  - 2.1.3 Preconditioning 38
  - 2.1.4 Conjugate gradient methods 40
- 2.2 Structural properties 46

		2.2.1	Quasi Newton	46

        2.2.2   Molecular dynamics 47

  2.3 Deriving thermodynamic properties 48

        2.3.1   Free energy surface 48

        2.3.2   Born-effective charges 50

3 S/PHI/nX   53

  3.1 Basis-set independent implementations 55

        3.1.1   Matrix notation 55

        3.1.2   *Bra-Ket* notation 56

        3.1.3   Programming languages 58

        3.1.4   BLAS/LAPACK interfaces 61

  3.2 The Dirac notation in S/PHI/nX 67

        3.2.1   Conventional approach 69

        3.2.2   Modular approach 69

        3.2.3   Object-oriented approach 76

        3.2.4   Example 88

  3.3 Class Hierarchy 88

        3.3.1   Electronic minimization 90

        3.3.2   Representing atomic structures 98

        3.3.3   Add-ons 102

  3.4 Comparison with VASP 103

  3.5 Conclusions 106

4 Applications   109

  4.1 Introduction 109

  4.2 Thermodynamic properties 110

        4.2.1   Convergence aspects 110

  4.3 Comparison with experiment 113

        4.3.1   Bulk properties at T=0 K 113

        4.3.2   Phonon spectra 115

        4.3.3   Thermal expansion 124

        4.3.4   Heat capacity 129

        4.3.5   Conclusions 132

5	**Conclusions and Outlook**	**135**
A	**Computational details**	**141**
	A.1 Pseudo potentials	141
	A.2 Convergence parameters	141
Bibliography		**143**

Introduction

The process of developing new materials is still mainly an empirical approach nowadays [1]. Typically, the materials are first being processed, then the structure and properties are investigated. Material properties are tuned by iterating this procedure with varying processing parameters. This approach is often very time-consuming and expensive. Due to advances in the physical and mathematical understanding of materials in combination with recent progresses in computer architectures, computer simulations are becoming a key technology to support future materials research and design. Since the computational approach allows not only to save costs but also to decrease the time to introduce new materials to the market, the new area of computational materials design (CMD) becomes an increasingly interesting research field in science and engineering. To address the fundamental questions of the emerging discipline great challenges in physics, mathematics/numerics, and computer sciences have to be addressed to improve efficiency and reliability and to enhance the predictive power of computations.

A wide range of methods and tools have been developed recently focusing on various single length and time scales. At the microscopic scale Density Functional Theory [2, 3] (DFT) has been proven to be reliable and efficient in predicting material properties. With modern computers and optimized DFT program packages it is possible to simulate systems consisting of 10^2 to 10^3 atoms. The theoretical description of larger systems requires further approximations, such as (semi-) empirical potentials [4]. For example, (Density Functional based) Tight-Binding (DF)TB [5, 6, 7] can treat systems up to 10^4 (DFTB) and even 10^7 (TB) atoms. In contrast to the above mentioned quantum-mechanical potentials also classical potentials, such as force fields (MM1 [8], MM2 [9], MM3 [10, 11, 12], MM4 [13, 14, 15, 16], AMBER [17, 18], CHARMM [19]), are often applied in particular for the simulation of biological systems.

Although these methods have been applied very successfully, there is a constant need for improving the existing algorithms. Increasing the performance as well as improving accuracy are always focal points in CMD method development. While in the past years highly optimized methods in single scales have been developed, the description of many material properties requires the consideration of various length- and time scales. Therefore, a new family of scale-bridging methods (e.g. multi-grids [20, 21], wavelets [22], heterogeneous multi-scale method [23], adaptive model refinement, coarse-grained simulations [24]) is currently in the focus of method development. Since computational materials design becomes also industrially applicable, it can be assumed that the improvement of existing algorithms and the connection of algorithms across the scales (multi-scale) as well as the combination of various physics disciplines (multi-physics) will become a more important research field in computational physics.

Besides the algorithmic work, method and code development are also critical elements of CMD. The computer architectures on which the program packages run have evolved significantly over the past years. Most of the applied methods in CMD (such as DFT) are computational very demanding and require program packages to be executed on high performance computers. Due to the remarkable running costs of such compute facilities, the optimization of CMD programs is crucial. The complexity of algorithmic optimization becomes clear considering that implementing algorithms in DFT in high level toolkits (such as Mathematica) can be accomplished within days or weeks while the development of an optimized CMD package applicable to realistic systems creates workloads of years [25, 26, 27, 28, 29, 30, 31, 32, 33, 34]! Since there is no approach available which is able to bridge the constantly broadening gap between computer science and computational physics, codes rarely reach peak performance. Method development in the field of high performance computing (HPC) requires thorough knowledge in both computer science and physics which opens a new inter-disciplinary field of research. Without such interdisciplinary work no modern simulation code can be developed nowadays.

In the scope of this work we contribute to closing the gap between computational physics and computer science. In our approach the physicists should be "screened" from the complex details of HPC. Therefore, a new physics programming meta-language has been developed which provides elements addressing the above mentioned requirements of CMD. (i) *State-of-the-art* computer science techniques have been applied or developed in this work to provide language elements for expressing algebraic expressions efficiently on modern high-performance computer architectures. (ii) In order to address the development of quantum mechanical algorithms which are crucial in CMD, the new meta-language supports the Dirac-formalism [35]. (iii) The approach is completed by an efficient way of expressing equations of motions to express structural algorithms such as molecular dynamics.

In this work we derive the details of the above mentioned meta-language. Based on this concept a new framework called S/PHI/nX has been developed in which the implementation of CMD algorithms becomes significantly simpler than in conventional approaches. We employ *state-of-the-art* methods from computer science to ensure peak performance of each language element. Within our framework physicists with only rudimentary programming skills are able to develop *efficient* CMD simulation programs.

As proof of concept we demonstrate the power of this approach by using the new framework for developing an efficient plane-wave DFT program package. It will be shown that the program is remarkably short with respect to the number of code lines which simplifies understanding and code maintenance. Since in this work we limit ourselves to deriving the concept, our program employs in the current version only norm-conserving pseudo potentials [36] and minimization methods optimized for semiconductors. Of course, the program package can easily be expanded by, e.g., modern basis-sets and more efficient minimizers.

The efficiency of the framework will be shown by means of calculations of realistic systems, namely the class of III-V semiconductors. Here the point of interest lies in both the execution speed as well as the obtained accuracy. A good test that requires very well converged parameters is the computation of thermodynamic properties like the heat capacity or the linear expansion coefficients from first-principles. In Ref. [37] the authors demonstrate how sensitive thermodynamic properties are with respect to the quality of the obtained forces from DFT calculations. Therefore, a study of these sensitive properties is an excellent test for the new framework.

From the physical and technological point of view the III-V semiconductors are most important. Electronic and opto-electronic devices, including blue and UV lasers, are made of these compounds. It is essential to understand their thermodynamic behavior during the growth process as well as their operation. Therefore, in this work we investigate the temperature dependencies of the most important thermodynamic properties.

III-V semiconductors can crystallize in the wurzite and zincblende phase. For many cubic III-V semiconductors the experimentally accessible data on these properties scatter significantly and reliable data obtained from first-principles are often not available. We, therefore, investigate the thermodynamic behavior of III-V semiconductors in the zincblende phase.

When computing properties within DFT a big degree of uncertainty arises from the exchange-correlation (XC) potential. The influence of XC is strongly material and system dependent. For cubic III-V semiconductors, however, there is no investigation clarifying the influence of the choice of the XC potential to the obtained thermodynamic properties. Thus, we compute the most important thermodynamic properties using both the local density approximation (LDA [38]) as well as the generalized-gradient approximation (GGA-PBE [39]).

The tetraedrically bound III-V semiconductors in the zincblende phase show a thermal anomaly at low temperatures. Up to a critical temperature the materials contract at rising temperatures. Only above the critical temperature these materials expand at increasing temperatures. While the underlying mechanism is essentially understood (e.g. Ref. [40]) the exact location of the critical point as well as the pronunciation of the anomaly is for some III-V semiconductors controversial. For example, in Ref. [41] GaP was reported not exhibiting the anomaly while according to Refs. [42] and [43] a tiny anomaly effect at low temperatures was observed. There is no *ab-initio* study available for this material in the cubic phase, neither is the influence of the XC potential to the pronunciation of the theoretically obtained anomaly investigated.

Reliable data of the heat capacity of some III-V systems are very scattered. If theoretical data from first-principles are available, they have been obtained only within LDA. We will extend the direct approach that recently has been applied to unary metallic systems [37] to binary systems in order to compute the phonon spectra, the temperature dependencies of the linear expansion coefficients as well as the heat capacities at constant pressure and constant volume.

Chapter 1

Theory

In this chapter we introduce the theoretical basis of this work and provide the required physical and mathematical/numerical formalism which is required to implement the efficient S/PHI/nX CMD library. On modern computer platforms some approaches are capable of describing very accurately systems containing only a few atoms while others can treat millions of atoms at a significantly lower accuracy [5, 6, 7]. Since a mixture or combination of various methods is often cumbersome, the development of a generalized framework is a major objective of this work. In the following sections our approach will be illustrated by means of a Hamiltonian employing norm-conserving pseudo potentials [36] as proof of concept. The wave functions will be represented in a plane-wave basis-set. Therefore, in this chapter the corresponding theoretical descriptions of the plane-wave basis as well as norm-conserving pseudo potentials are discussed. For the design of a *general* framework it is, however, important that the concept can be easily extended to describe alternative valence/core partitioning. We, therefore, briefly sketch various alternative methods ranging from APW [44] to FP-LAPW+lo [45, 46] and PAW [31]. Although they will not be implemented in the scope of this work, the framework will prepare interfaces and concepts which allow a straightforward implementation of these methods. Besides an accurate but computationally demanding *ab-initio* description of systems, the framework should be applicable for larger systems. Here, the number of atoms that can be computed in reasonable times can be traded with lower accuracy using, e.g., (semi-)empirical methods. In the end of this chapter tight-binding methods, structural optimization schemes, and molecular dynamics are briefly sketched.

1.1 The Many-body problem

Calculations on the atomic scale focus on the description of atoms, molecules, and crystals. In quantum mechanics a system of n_{el} electrons and N_I ions can be represented by a Hamilton operator $\hat{H}$

$$\hat{H} = \hat{T}_e + \hat{T}_I + \hat{v}_{ee} + \hat{v}_{eI} + \hat{v}_{II} \tag{1.1}$$

with $\hat{T}_e$ and $\hat{T}_I$ being the kinetic energy of the electronic and ionic subsystem, respectively. The potential contribution $\hat{v}_{ee}$ and $\hat{v}_{II}$ denote the repulsive electron-electron and ion-ion Coulomb interactions, respectively. The remaining potential contribution $\hat{v}_{eI}$ is the attractive Coulomb interaction between the n_{el} electrons and N_I ions. With M_I being the mass of the Ith ion and Z_I the corresponding atomic number the contributions to the Hamiltonian become[1]

$$\hat{H} = -\sum_i^{n_{el}} \frac{1}{2}\Delta_i - \sum_I^{N_I} \frac{1}{2M_I}\Delta_I + \sum_i^{n_{el}}\sum_{j>i}^{n_{el}} \frac{1}{|\mathbf{r}_i - \mathbf{r}_j|} - \sum_i^{n_{el}}\sum_I^{N_I} \frac{Z_I}{|\mathbf{r}_i - \tau_I|} + \sum_I^{N_I}\sum_{J>I}^{N_I} \frac{Z_I Z_J}{|\tau_I - \tau_J|}. \quad (1.2)$$

The symbol $\mathbf{r}_i$ refers to the position of the i-th electron whereas the location of the I-th ion is specified by τ_I.

The time-independent Schrödinger [47] equation provides a quantum mechanical description of the system

$$\hat{H}|\Psi^{\mathrm{MB}}\rangle = E|\Psi^{\mathrm{MB}}\rangle. \quad (1.3)$$

Here $|\Psi^{\mathrm{MB}}\rangle$ denotes the many-body wave function of the nuclei and the electrons. The symbol E is the total energy of the system. Solving the Schrödinger equation (1.3) gives access to all (static) properties of the many-body problem.

The variational principle With the Rayleigh-Ritz variational principle [48, 49] a basis for numerical solutions of the Schrödinger equation (1.3) is available. According to this principle, the expectation value of a Hamiltonian calculated for *any* trail wave function Ψ_{trial} is always greater or equal to the ground state energy

$$E^0 \leq E[\Psi^{\mathrm{MB}}] = \frac{\langle \Psi^{\mathrm{MB}}|\hat{H}|\Psi^{\mathrm{MB}}\rangle}{\langle \Psi^{\mathrm{MB}}|\Psi^{\mathrm{MB}}\rangle} \quad (1.4)$$

$$\text{or } E^0 = \min_{\{\Psi^{\mathrm{MB}}\}} E[\Psi^{\mathrm{MB}}]. \quad (1.5)$$

With the variational principle the Schrödinger equation (1.3) can be reformulated

$$\delta\left(\langle \Psi^{\mathrm{MB}}|\hat{H}|\Psi^{\mathrm{MB}}\rangle - E\langle \Psi^{\mathrm{MB}}|\Psi^{\mathrm{MB}}\rangle\right) = 0. \quad (1.6)$$

This equation gives rise to a way of solving the Schrödinger equation iteratively. The total energy E is a Lagrange multiplier connected to the normalization of the solution Ψ^{MB}.

1.1.1 Born-Oppenheimer approximation

For realistic systems the Schrödinger equation (1.3) has 10^{23} degrees of freedom and all particles interact with each other according to Eq. (1.2). The complexity of the quantum mechanical many-body problem can be reduced by separating the electronic motion from that of the ions. Such an approximation can be justified with the relatively huge mass of the ions compared to that of the electrons ($\frac{m_e}{M_I} \ll 1$): The neglected energy contributions are by $\frac{m}{M_I}$ and $(\frac{m}{M_I})^{\frac{1}{2}}$ smaller than the electronic energy (see e.g. Ref. [45]) and are, therefore, much smaller than the distances between electronic levels. Approximately, the electrons follow the (slower) ions adiabatically. Therefore, this approximation is called adiabatic approximation, also known as Born-Oppenheimer (BO) approximation[2] [51].

---
[1] Unless noted otherwise, we use atomic units throughout this work, i.e., $m_e = c = \hbar = \frac{e^2}{4\pi\epsilon_0} = 1$.
[2] The Born-Oppenheimer approximation fails, however, for systems where the nuclei movement is relatively fast with respect to the electronic system [50], e.g., highly excited rotational states of molecules. Here the fast molecular movements do not allow the electronic system to follow adiabatically!

By keeping the positions of the nuclei $\tau = (\tau_1, \ldots, \tau_{N_i})$ fixed, the equations of motion of the electrons at $\mathbf{r} = (\mathbf{r}_1, \ldots, \mathbf{r}_n)$ decouple from those of the nuclei. In this separation ansatz the nuclei merely modulate the electronic wave functions and the total wave function can be expressed as

$$\Psi^{MB}(\mathbf{r}, \tau) = \Psi_I^{BO}(\tau)\Psi_e^{BO}(\mathbf{r}, \tau). \tag{1.7}$$

Ψ_e^{BO} refers to the solution of the Schrödinger equation with neglected kinetic energy of the nuclei $\hat{T}_I$. Ψ_I^{BO} is the wave function of the ions. The total energy values $E(\{\tau_I\})$ at a fixed set of ionic positions form the so-called Born-Oppenheimer surface.

1.1.2 Electron-electron interaction

Even after applying the Born-Oppenheimer approximation the quantum mechanical many-body problem is still far too complex, mainly due to the remaining electron-electron interaction determined by the Pauli principle and the Coulomb interaction between the electrons.

The first approach to cope with this difficulty was suggested by **Hartree** [52]. Here, the many-body wave function $|\Psi^{MB}\rangle$ is approximated by a product of single-particle wave functions $|\psi_i\rangle$. Each single particle is moving in an averaged self-consistent potential of all single particles. $|\psi_i\rangle$ satisfies the single-particle Schrödinger equation. The Hartree approach does not incorporate the Pauli principle. In order to do so in the **Hartree-Fock** approach [53, 54] the wave function is approximated by an anti-symmetrized product of n orthonormal spin orbitals $\psi_i^{HF}(\mathbf{r})$. Each orbital is constructed from a spatial orbital $\phi_i(\mathbf{r})$ and a spin function $\chi(\sigma)$. The Hartree-Fock wave function Ψ_i^{HF} is constructed from a Slater determinant of the ψ_i^{HF} according to

$$\Psi_i^{HF} = \frac{1}{\sqrt{n!}} |\psi_i^{HF}| \qquad \psi_i^{HF} = \phi_i(\mathbf{r})\chi(\sigma) \tag{1.8}$$

with the orthonormality constraint

$$\langle \psi_i^{HF} | \psi_j^{HF} \rangle = \delta_{ij}.$$

The Hartree-Fock equations read

$$\int \left(-\frac{1}{2}\Delta\delta(\mathbf{r}-\mathbf{r}') + v_H(\mathbf{r})\delta(\mathbf{r}-\mathbf{r}') + v_{el}(\mathbf{r})\delta(\mathbf{r}-\mathbf{r}') + v_X(\mathbf{r},\mathbf{r}') \right) \psi_i^{HF} d\mathbf{r}' = \sum_{i \neq j} \varepsilon_{ij} \psi_i^{HF}(\mathbf{r}).$$

The electron-electron interaction is referred to as the Hartree term $v_H(\mathbf{x})$. The new non-local term $v_X(\mathbf{r}, \mathbf{r}')$ is called exchange. It describes the energy gain due to the anti-symmetrization of the wave function when two electrons with originally equal spin σ reduce their Coulomb energy by flipping the spin of one electron and occupying the same orbital. The Hartree-Fock equations can be solved in a self-consistent field (SCF) calculation.

By combining Slater determinants (so-called "multi configuration") the above picture can be improved. In the **configuration interaction** [55] (CI) such linear combinations of Slater determinants of many configurations are used to approximate the many-body wave function. CI corrects the lack of correlation effects in the Hartree-Fock approach and leads theoretically to an exact many-body wavefunction. However, the computational effort in CI scales exponentially with the system size. Hence, CI can be presently applied to very small systems only (~ 20 atoms) [1].

In the **Thomas-Fermi model** [56, 57, 58] statistical considerations lead to an approximation of the electron distribution in an atom. The first assumption is that the electrons are uniformly distributed in the phase space per h^3 volume box with h being the Planck constant. The second assumption is the existence of an effective potential which is determined by the charge of the nuclei and the electrons. From these assumptions an energy expression depending only of the electron charge density can be derived. The kinetic energy is being described very poorly in this model. The Thomas-Fermi model can be considered as the origin of Density Functional Theory.

1.2 Density functional theory

In this section we provide a brief overview about Density Functional Theory (DFT) which is one of the main focal points of this work. By providing two fundamental theorems Hohenberg and Kohn [2] accomplished to form an exact theory that can be used for numerical calculations of realistic systems. The first theorem shows that instead of the complex wave function the much simpler charge density can act as the key entity when calculating the electronic ground state. The second theorem provides a way how an actual minimization scheme can be realized. These two theorems are the foundation of DFT.

1.2.1 Kohn-Sham formalism

First Hohenberg Kohn theorem Both the ground state energy E_0 as well as the many body wave function Ψ^{MB} of an electronic system which is specified by an Hamilton operator as in Eq. (1.2) can be obtained by minimizing [59] the energy functional $E[\Psi]$

$$E_0 \leq E[\Psi^{MB}] = \frac{\langle \Psi^{MB}|\hat{H}|\Psi^{MB}\rangle}{\langle \Psi^{MB}|\Psi^{MB}\rangle}. \tag{1.9}$$

The electronic system is, of course, determined by the underlying atomic structure given by an external potential $v(\mathbf{R})$ as well as the number of electrons n. Hence, the two entities $v(\mathbf{R})$ and n alone determine entirely the Hamiltonian of a system and thus, the electronic ground state energy.

The previous statement can be formulated even stronger because both $v(\mathbf{R})$ and n are given by the density (and a trivial constant c)

$$v = v[\varrho(\mathbf{R})] + c, \tag{1.10}$$

$$n[\varrho(\mathbf{R})] = \int \varrho(\mathbf{R})d\mathbf{R}. \tag{1.11}$$

The external potential - and hence the energy and the ground state wave function of an electronic system - is entirely determined, within an additive constant[3], by the electron density $\varrho(\mathbf{R})$ [50].

This theorem is the **first Hohenberg Kohn theorem**.

---
[3]The here involved constant is usually chosen such that v vanishes at $\mathbf{R} \to \infty$.

Second Hohenberg Kohn theorem The energy E_v for a given external potential $v(\mathbf{R})$ can be written as

$$E_v[\varrho] = \underbrace{T[\varrho] + v_{ee}[\varrho]}_{F_{\text{HK}}[\varrho]} + v_{eI}[\varrho] \quad (1.12)$$

$$= F_{\text{HK}}[\varrho] + \int \varrho(\mathbf{R})v(\mathbf{R})d\mathbf{R}. \quad (1.13)$$

The introduced term F_{HK} is called Hohenberg-Kohn functional and will be discussed in detail below. In analogy to the variational principle for wave functions (Eq. 1.4) the **second Hohenberg-Kohn theorem** provides an energy variational principle for densities:

The functional $E_v[\varrho]$ of Eq. (1.12) becomes minimal at the correct ground state density. That minimal value corresponds to the ground state energy value E_0 [50].

That means that for any (positive) trial density ϱ_{trial}, which fulfills $\int \varrho_{\text{trial}}(\mathbf{R})d\mathbf{R} = n$, the inequality

$$E_0 \leq E_v[\varrho_{\text{trial}}] \quad (1.14)$$

holds.

According to the second Hohenberg-Kohn theorem the problem of determining the ground state entities E_0, ϱ_0, and Ψ_0 for a given external potential is nothing but minimizing the energy functional $E_v[\varrho]$ of Eq. (1.12) with respect to the density. With the constraint of keeping the number of electrons constant $\int \varrho(\mathbf{R})d\mathbf{R} = n$ the following equation has to be minimized

$$\delta \left(E_v[\varrho] - \mu (\int \varrho(\mathbf{R})d\mathbf{R} - n) \right) = 0, \quad (1.15)$$

which gives with Eq. (1.12)

$$\mu = \frac{\delta E_v[\varrho(\mathbf{R})]}{\delta \varrho(\mathbf{R})} = v(\mathbf{R}) + \frac{\delta F_{\text{HK}}[\varrho]}{\delta \varrho(\mathbf{R})}. \quad (1.16)$$

The Lagrange multiplier μ which ensures the previously mentioned electron conservation can be identified with the chemical potential of the electrons.

1.2.2 Kohn-Sham equations

In order to apply the DFT formalism to realistic systems an explicit form of the Hohenberg-Kohn density functional $F_{\text{HK}}[\varrho]$ is required. An approximation which obtains rather accurate results has been introduced in the Kohn-Sham method [3] which shall be briefly discussed in this section.

Introduction of Kohn-Sham orbitals Kohn and Sham suggested an indirect method to compute the unknown functional $F_{\text{HK}}[\varrho]$. They introduced Kohn-Sham orbitals Ψ_i such that the kinetic energy $T[\varrho]$ can be determined to a reasonable accuracy leaving a small residual correction which is handled separately. Therefore, an auxiliary system of *non-interacting* particles was introduced with the single-particle kinetic energy T_s and the local single-particle potential v_s such that the single-particle ground state densities of the interacting and non-interacting systems are equal. The functional $F_{\text{HK}}[\varrho]$ can then be expressed as follows

$$F_{\text{HK}}[\varrho] = T[\varrho] + v_{\text{ee}}[\varrho] \tag{1.17}$$
$$= T_s[\varrho] + J[\varrho] + \underbrace{(T[\varrho] - T_s[\varrho]) + (v_{\text{ee}}[\varrho] - J[\varrho])}_{E_{\text{xc}}[\varrho]} \tag{1.18}$$
$$= T_s[\varrho] + J[\varrho] + E_{\text{xc}}[\varrho]. \tag{1.19}$$

Here, $J[\varrho]$ symbolizes the classical electrostatic repulsion of the electrons. E_{xc} is called the **exchange-correlation energy** consisting of [50]

- the difference between the kinetic energy and the single-particle kinetic energy $(T - T_s)$ and
- the non-classical part of electron-electron interaction $(v_{\text{ee}} - J)$.

Note that after rewriting the functional $F[\varrho]$ only the still unknown expression $E_{\text{xc}}[\varrho]$ contains terms of the interacting electron system whereas T_s and J describe the system of non-interacting electrons.
The kinetic energy of the non-interacting system in terms of the n lowest single-particle orbitals is

$$T_s[\varrho] = \sum_i^n \langle \Psi_i | -\frac{1}{2}\nabla^2 | \Psi_i \rangle. \tag{1.20}$$

The single-particle ground state density is computed from

$$\varrho(\mathbf{R}) = \sum_i^n |\Psi_i(\mathbf{R})|^2. \tag{1.21}$$

The above expressions for the kinetic energy and for the density hold only as long as the Kohn-Sham orbitals fulfill the orthonormalization constraint

$$\delta_{ij} = \langle \Psi_i | \Psi_j \rangle. \tag{1.22}$$

The energy functional can now be rewritten in terms of the n Kohn-Sham orbitals

$$E[\varrho] = F_{\text{HK}}[\varrho] + \int v(\mathbf{R})\varrho(\mathbf{R})d\mathbf{R}$$
$$= T_s[\varrho] + J[\varrho] + E_{\text{xc}}[\varrho] + \int v(\mathbf{R})\varrho(\mathbf{R})d\mathbf{R}$$
$$= \sum_i^n \langle \Psi_i | T_s | \Psi_i \rangle + J[\varrho] + E_{\text{xc}}[\varrho] + \int v(\mathbf{R})\varrho(\mathbf{R})d\mathbf{R}. \tag{1.23}$$

The Euler equation (1.16) belonging to the above energy function can be expressed as follows

$$\mu = \frac{\delta E[\varrho(\mathbf{R})]}{\delta \varrho(\mathbf{R})} = \frac{\delta T_s[\varrho]}{\delta \varrho(\mathbf{R})} + \underbrace{\frac{\delta J[\varrho]}{\delta \varrho(\mathbf{R})} + \frac{\delta E_{\text{xc}}[\varrho]}{\delta \varrho(\mathbf{R})} + v(\mathbf{R})}_{v_{\text{eff}}(\mathbf{R})} \qquad \frac{\delta J[\varrho]}{\delta \varrho(\mathbf{R})} = v_{\text{H}} = \int \frac{\varrho(\mathbf{R}')}{|\mathbf{R}-\mathbf{R}'|}d\mathbf{R}'$$

$$\frac{\delta E_{\text{xc}}[\varrho]}{\delta \varrho(\mathbf{R})} = v_{\text{xc}}$$

$$\mu = \frac{\delta T_s}{\delta \varrho(\mathbf{R})} + v_{\text{eff}}(\mathbf{R}). \tag{1.24}$$
$$v_{\text{eff}}(\mathbf{R}) = v_{\text{H}} + v_{\text{xc}} + v(\mathbf{R}) \tag{1.25}$$

Note that so far this equation is nothing but a rearrangement of the Euler equation (1.16).

Search of the energy minimum Based on the rewritten energy functional Eq. (1.23) it can be concluded that the variational search of the minimum of $E[\varrho]$ can be performed in the space of the single particle Kohn-Sham orbitals $\{\Psi_i\}$. Therefore, a functional Ω of the Kohn-Sham orbitals can be defined including Lagrange multipliers λ_{ij} to enforce orthonormalization (Eq. (1.22))

$$\Omega[\{\Psi_i\}] = E[\varrho] - \sum_i \sum_j \lambda_{ij} \left(\langle \Psi_i | \Psi_j \rangle - \delta_{ij} \right).$$

Finding the minimum of $E[\varrho]$ in the space of the Kohn-Sham orbitals implies

$$\delta\Omega[\{\Psi_i\}] \stackrel{!}{=} 0 \;=\; \underbrace{\frac{\delta E[\varrho]}{\delta \varrho}}_{\text{Eq. (1.24)}} \underbrace{\frac{\delta \varrho}{\delta \langle \Psi_i |}}_{|\Psi_i\rangle} - \sum_j \lambda_{ij} |\Psi_i\rangle \quad (1.26)$$

$$= (T_\text{s} + v_\text{eff})|\Psi_i\rangle - \sum_j \lambda_{ij}|\Psi_i\rangle \quad (1.27)$$

$$\Longrightarrow (T_\text{s} + v_\text{eff})|\Psi_i\rangle = \sum_j \lambda_{ij}|\Psi_i\rangle. \quad (1.28)$$

The λ_{ij} represent a hermitian matrix which can be diagonalized by applying an uniform transformation to the Kohn-Sham orbitals $|\Psi_i\rangle$. The Hamiltonian as well as the density are invariant with respect to the uniform transformation. Hence, the above equation can be reformulated as follows

$$\hat{H}_\text{eff}|\Psi'_i\rangle = (T_\text{s} + v_\text{eff})|\Psi'_i\rangle = \varepsilon_i|\Psi'_i\rangle. \quad (1.29)$$

By priming the Kohn-Sham orbitals it should be emphasized that the uniform transformation has been applied to Ψ_i. The $|\Psi'\rangle$ are now eigenfunctions of $\hat{H}_\text{eff}$. In the following the prime will be omitted.

Solving the Kohn-Sham equations The ground state of the interacting system can be found by solving the equations

$$\left(-\frac{1}{2}\Delta + v_\text{eff}(\mathbf{R}) \right) \Psi_i = \varepsilon_i \Psi_i, \qquad i = 1, 2, \ldots, n. \quad (1.30)$$

and

$$\varrho(\mathbf{R}) = \sum_{i=1}^n |\Psi_i(\mathbf{R})|^2. \quad (1.31)$$

As v_eff depends on $\varrho(\mathbf{R})$ via Eq. (1.24), the equations (1.25), (1.30), and (1.31) have to be solved self-consistently. That means that the non-interacting electrons are moving in the effective self-consistent field of all electrons. Such an approach is called self-consistent field (SCF) calculation. These three equations are known as the **Kohn-Sham equations** and build up the backbone of the Kohn-Sham density functional theory

$$\begin{align}
\text{I:} \quad & v_{\text{eff}}(\mathbf{R}) = v(\mathbf{R}) + v_{\text{H}}(\mathbf{R}) + v_{\text{xc}}(\mathbf{R}) \\
\text{II:} \quad & \left(-\tfrac{1}{2}\Delta + v_{\text{eff}}(\mathbf{R})\right)\Psi_i = \varepsilon_i \Psi_i \\
\text{III:} \quad & \varrho(\mathbf{R}) = \sum_{i=1}^{n} |\Psi_i|^2.
\end{align} \quad (1.32)$$

Conclusion and interpretation of the Kohn-Sham equations The Kohn-Sham method transforms the many-body problem of interacting electrons into an effective single-particle problem by introducing n Kohn-Sham orbitals. In DFT the many-body problem is not specified by the complex many-body wave function $\Psi^{\text{MB}}(\mathbf{r}_1, \mathbf{r}_2, \mathbf{r}_3, \ldots, \mathbf{r}_n)$ with $3n$ coordinates any longer. Instead the electron density ϱ with only three spatial dimensions becomes the key entity.

In contrast to the Thomas-Fermi method the kinetic energy of the non-interacting system is correctly obtained. In return the computational effort is increased because the single equation to obtain the total density becomes a system of n equations which has to be solved. The Kohn-Sham equations are reminiscent to the previously mentioned Hartree equations. A major advantage over the Hartree method is that v_{eff} provides a way to incorporate exchange-correlation effects. Solving the Kohn-Sham equations is computationally less demanding than the Hartree-Fock equations, mainly due to the non-local Fock operator. Compared to CI the Kohn-Sham equations are dramatically simpler to evaluate.

Up to now the functional $E_{\text{xc}}[\varrho]$ in Eq. (1.23) is still undefined. Since E_{xc} contains terms of the interacting system it is clear that only approximate expressions for this term can be found. It is also obvious that the way of such an approximation strongly depends on how the density varies in space and is hence, system dependent. The following section is, therefore, dedicated to this issue.

1.2.3 XC functional

In order to specify the Kohn-Sham equations (1.32) an explicit form of the exchange correlation contribution is still missing. The search for approximations providing high accuracies of the exchange correlation functional is up to today one of the greatest challenges in DFT.

Local density approximation (LDA) The first and simplest approach to find an approximate expression for an exchange correlation functional is to start from a uniform electron distribution.

$$E_{\text{xc}}^{\text{LDA}}[\varrho] = \int \varrho(\mathbf{R}) \epsilon_{\text{xc}}^{\text{hom}}[\varrho] d\mathbf{R}. \quad (1.33)$$

where $\epsilon_{\text{xc}}^{\text{hom}}$ is the exchange correlation energy *per particle* of the homogeneously distributed electron gas. It can be divided into an exchange and a correlation part

$$\epsilon_{\text{xc}}^{\text{hom}}[\varrho] = \epsilon_{\text{x}}^{\text{hom}}[\varrho] + \epsilon_{\text{c}}^{\text{hom}}[\varrho]. \quad (1.34)$$

The Thomas-Fermi model provides an expression for the kinetic energy T_s as well as the exchange energy ε_{x} of the uniform electron gas. The exchange energy per particle in the Thomas-Fermi model reads

$$\epsilon_{\text{x}}^{\text{TF}}[\varrho] = -\frac{3}{4}\left(\frac{3}{\pi}\right)^{\frac{1}{3}} \varrho(\mathbf{R})^{\frac{1}{3}}.$$

The correlation part ϵ_c^{hom} must contain the remaining unknown contributions to E_{xc} namely, the non-classical part of the electron-electron interaction ($v_{\text{ee}} - J$) as well as the difference $T - T_s$. An analytic expression of the ϱ dependence of ϵ_c is not available. However, due to quantum Monte Carlo calculations by Ceperley and Alder [38] an interpolation formula to ϵ_c is at hand.

Accordingly, the exchange correlation potential v_{xc} from Eq. (1.32) becomes

$$v_{\text{xc}}^{\text{LDA}}(\mathbf{R}) = \frac{\delta E_{\text{xc}}^{\text{LDA}}}{\delta \varrho(\mathbf{R})} = \epsilon_{\text{xc}}^{\text{hom}} \varrho(\mathbf{R}) + \epsilon_{\text{xc}}^{\text{hom}} \frac{\delta \epsilon_{\text{xc}}^{\text{hom}}}{\delta \varrho(\mathbf{R})}. \tag{1.35}$$

LDA is a good approximation for system with slowly varying electron densities such as many bulk systems. For systems such as atoms and molecules that show inhomogeneous densities, however, LDA may become too inaccurate. In comparison with the experiment LDA often predicts too small lattice constants and bond distances (overbinding). Binding and cohesive energies are usually too large [60].

Generalized Gradient Approximation (GGA) For systems in which the charge densities cannot be simply approximated by an uniform electron gas the generalized gradient approximation (GGA) can be used. Commonly, besides the electronic charge density also the gradient of the density is taken into account. Such an approximation is referred to as generalized gradient approximation

$$E_{\text{xc}}^{\text{GGA}}[\varrho] = \int \varrho(\mathbf{R}) \epsilon(\varrho, \nabla \varrho) d\mathbf{R}. \tag{1.36}$$

In general, the errors introduced by the exchange-correlation functional cannot be quantified. For many systems GGA was found to correct the overbinding problem of LDA. The cohesive energies are often significantly improved by applying GGA [60].

A major focus of this work is the investigation of thermodynamic properties of III-V semiconductors. In order to estimate the uncertainty arising from XC, in this work all calculations will be performed with both LDA and GGA-PBE [39]. Therefore, both LDA and GGA-PBE will be implemented in the S/PHI/nX framework.

1.3 Periodic boundary conditions

When applying the Kohn-Sham formalism to periodic systems such as crystals, an infinite number of ions have to be treated. Furthermore, the wave functions extend over the entire space. Hence, an infinite basis-set would be required. For systems with periodic boundary conditions such as crystals the dimensionality of the many-electron system can be drastically reduced when employing translational symmetry.

The periodic atomic structure of a crystal creates an periodic external potential $v^{\text{ext}}(\mathbf{R})$ in which the electrons move

$$v_{\text{ext}}(\mathbf{R} + \mathbf{R}_{\text{lat}}) = v_{\text{ext}}(\mathbf{R}). \tag{1.37}$$

The periodicity is given by the lattice vectors

$$\mathbf{R}_{\text{lat}} = \sum_{i=1}^{3} n_i \mathbf{a}_i, \quad \Omega = |\mathbf{a}_1 \cdot \mathbf{a}_2 \times \mathbf{a}_3|. \tag{1.38}$$

The three lattice vectors $\mathbf{a}_i$ are the lattice vectors of the primitive unit cell with the volume Ω. The n_i denote integer numbers. The translational symmetry of $v_{\text{ext}}(\mathbf{R})$ suggests that also the Hamiltonian underlies the same translation symmetry. According to Bloch's theorem, the wave functions of such a Hamiltonian with translational invariance can be factorized in a cell periodic part $f_i(\mathbf{R})$ and a wave-like part (phase factor) (see Ref. [47])

$$\Psi_{i\mathbf{k}}(\mathbf{R}) = e^{i\mathbf{k}\cdot\mathbf{R}}f_i(\mathbf{R}), \qquad f_i(\mathbf{R}+\mathbf{R}_{\text{lat}}) = f_i(\mathbf{R}). \tag{1.39}$$

The indices i denote the band index and $\mathbf{k}$ point which lies in the first Brillouin zone, respectively.

1.4 Integration over the Brillouin zone

The computation of expectation values requires an integration over the Brillouin zone. In a bulk crystal all occupied states i at each of the infinite $\mathbf{k}$ points contribute to the density $\varrho(\mathbf{R})$ and thus, to the potential $v(\mathbf{R})$. Hence, when computing the potential an infinite number of calculations is required. Wave functions are smooth in reciprocal space and are almost identical at $\mathbf{k}$ points which are close to each other [61]. Therefore, the wave functions in a region of $\mathbf{k}$ points can be approximately expressed by a single wave function at a *representative single* $\mathbf{k}$ point. This allows to consider electronic states at a *finite* number of $\mathbf{k}$ points. The integral over the Brillouin zone can be replaced by a discrete sum over a chosen $\mathbf{k}$-point mesh

$$\int_{\text{BZ}} d\mathbf{k} \to \sum_{\mathbf{k}} \omega_{\mathbf{k}} \Delta\Omega. \tag{1.40}$$

The $\omega_{\mathbf{k}}$ are weight factors which fulfill the conservation law

$$\sum_{\mathbf{k}} \omega_{\mathbf{k}} = 1. \tag{1.41}$$

The choice of $\mathbf{k}$-points used for sampling the Brillouin zone determines the quality of the obtained SCF results. There are various methods known to sample the $\mathbf{k}$-space in order to integrate continuous functions over the Brillouin zone [31, 62, 63, 64, 65, 66]. One of the most often applied techniques is the special-point scheme of Monkhorst and Pack [62] which will be implemented in the plane-wave framework S/PHI/nX (see Fig. 1.1).

The Monkhorst-Pack scheme is very successful for total energy calculations of semiconductors and insulators. However, at a first glimpse it fails at T=0 K when performing total energy calculations of metallic systems because the function to be integrated becomes discontinuous at the Fermi edge. In order to cope with this problem occupation numbers $f_{i\sigma\mathbf{k}}^{\text{occ}}$ can be introduced when computing the electronic charge density

$$\varrho(\mathbf{R}) = \sum_{i\sigma\mathbf{k}} \omega_{\mathbf{k}} f_{i\sigma\mathbf{k}}^{\text{occ}} |\Psi_{i\sigma\mathbf{k}}|^2 \tag{1.42}$$

which can be obtained according to the Fermi function

$$f_{i\sigma\mathbf{k}}^{\text{occ}} = \frac{1}{\exp(\frac{\varepsilon_{i\sigma\mathbf{k}}-\varepsilon_F}{k_B T})+1}. \tag{1.43}$$

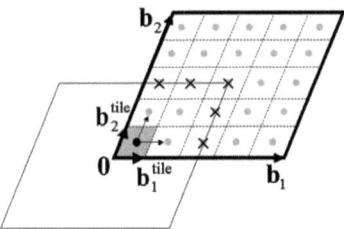

Figure 1.1: Schematic representation of the Monkhorst-Pack scheme for the case of a 2D lattice. The parallelogram centered around the origin is the conventional Brillouin zone whereas the bold one indicates the technical implementation in the S/PHI/nX code. We span the Brillouin zone by the reciprocal lattice vectors $\mathbf{b}_i$. The Brillouin zone is subdivided into identical small tiles spanned by the vectors $\mathbf{b}_i^{\text{tile}}$. A set of generating k-points (here one generating k-point at $(\frac{1}{2}\frac{1}{2})$) is placed into each tile to generate all special k-points (gray circles). Points along the edges of the conventional Brillouin zone (marked with 'x') should be avoided when constructing the Monkhorst-Pack k-point mesh. The set of necessary k-points can be reduced by applying crystal symmetries.

In the Fermi function the $\varepsilon_{i\sigma\mathbf{k}}$ are the one-particle energies, ε_F is the Fermi energy, k_B is the Boltzmann constant, and T is the temperature.

Super cell approach The Bloch-theorem can be applied to periodic systems only. Calculations of surfaces, for example, would require an infinite number of basis-set functions perpendicular to the surface plane. Hence, numerical calculations would be impossible.

However, by introducing super cells which mimic the non-periodicity and repeating them periodically, Bloch's theorem can also be applied to such systems. In Fig. (1.2) a sketch of the super cells of a defect calculation and a surface calculation is presented. The super cell has to be chosen large enough that the feature (defect, vacuum, slab) causing the break of the translation symmetry is nearly decoupled from its images.

Plane-wave representation of KS orbitals The methods and techniques which will be derived in this work will be demonstrated by means of a pseudo potential plane-wave library. Therefore, the KS orbitals will be represented in a plane-wave basis-set. The expressions contributing to the potential v_eff and the total energy E_tot that are required for the implementation of our framework will be presented in the following.

The cell periodic part f can be expanded using a basis-set $e^{i\mathbf{G}\cdot\mathbf{R}}$ with of *discrete* set of plane-waves

$$f_i(\mathbf{R}) = \sum_{\mathbf{G}} c_i(\mathbf{G}) e^{i\mathbf{G}\cdot\mathbf{R}}. \tag{1.44}$$

The expansion coefficients are labeled c_i. The reciprocal lattice vectors $\mathbf{G}$ that fulfill $\mathbf{G}\cdot\mathbf{R}_\text{lat} = 2\pi m$ with m being any integer number. Using Bloch's theorem the wave function reads finally

$$\Psi_{i\mathbf{k}}(\mathbf{R}) = \sum_{\mathbf{G}} c_{i\mathbf{k}}(\mathbf{G}) e^{i(\mathbf{G}+\mathbf{k})\cdot\mathbf{R}}. \tag{1.45}$$

The application of Bloch's theorem transforms the problem of describing an *infinite* number of wave functions expanded over the *infinite* space to an *infinite* number of wave functions defined only in the first unit cell

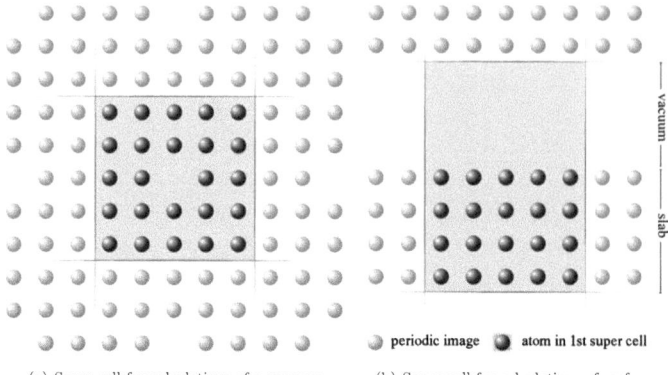

(a) Super cell for calculations of a vacency (b) Super cell for calculations of surfaces

Figure 1.2: Illustration of the super cell approach. The first unit cell (darker, blue balls) is repeated along all spatial directions. (a) Setup of a super cell to compute a vacancy defect in a bulk solid. The size of the 1st super cell has to be chosen large enough in order to decouple the vacancy from its image defects. (b) The super cell approach can also be applied to surface calculations. In this case the vacuum region has to be big enough to decouple the two surfaces from each other. On the other hand the slab region needs to contain enough atomic layers to decouple the two vacuum regions from each other. Otherwise a thin film would be modeled.

of the crystal. On the other hand, however, Bloch's theorem makes the wave functions to be given at an *infinite* set of **k** points.

Eq. (1.45) is an expansion of the Kohn-Sham orbitals in a plane-wave basis set. In this basis-set the Kohn-Sham equation can be written as [61]

$$\sum_{\mathbf{G}'} H_{\mathbf{G}+\mathbf{k},\mathbf{G}'+\mathbf{k}} c_{i\mathbf{k}}(\mathbf{G}') = \varepsilon_{i\mathbf{k}} c_{i\mathbf{k}}(\mathbf{G}) \qquad (1.46)$$

$$\sum_{\mathbf{G}'} \left(|\mathbf{G}+\mathbf{k}|^2 \delta_{\mathbf{G}\mathbf{G}'} + v_{\text{ion}}(\mathbf{G}-\mathbf{G}') + v_{\text{H}}(\mathbf{G}-\mathbf{G}') + v_{\text{xc}}(\mathbf{G}-\mathbf{G}') \right) c_{i\mathbf{k}}(\mathbf{G}') = \varepsilon_{i\mathbf{k}} c_{i\mathbf{k}}(\mathbf{G}). \qquad (1.47)$$

In this representation the kinetic energy is diagonal in **G** space, the contributions to the effective potential are described in terms of their Fourier transforms using

$$\langle \mathbf{R} | \mathbf{G}+\mathbf{k} \rangle = \Omega \sum_{\mathbf{G}} e^{+i(\mathbf{G}+\mathbf{k}) \cdot \mathbf{R}}. \qquad (1.48)$$

The expansion coefficients $c_{i\mathbf{k}}(\mathbf{G})$ can be determined by diagonalizing the Hamiltonian matrix $H_{\mathbf{G}+\mathbf{k},\mathbf{G}'+\mathbf{k}}$. In principle, an infinite basis-set is required to expand the wave functions $\Psi_{i\mathbf{k}}$. Hence, also the Hamilton matrix $H_{\mathbf{G}+\mathbf{k},\mathbf{G}'+\mathbf{k}}$ would have an infinite size. Typically, only the coefficients $c_{i\mathbf{k}}(\mathbf{G})$ with small kinetic energy contributions are more important than those with large kinetic energies [61]. Therefore, the plane-wave basis-set can be truncated at a certain energy cut-off

$$E_{\text{cut}} = \frac{1}{2} |\mathbf{G}+\mathbf{k}|^2_{\text{max}} \qquad (1.49)$$

which defines the highest kinetic energy of the basis functions as well as the shortest wave length

$$\lambda_{\min} = \frac{2\pi}{G_{\max}}. \tag{1.50}$$

Hence, the basis-set convergence can be systematically improved by reducing the wave length which is equivalent to increasing the energy cut-off. In this work the cut-off is given in Rydberg[4].

With the above given Hamiltonian (Eq. (1.46)) the **total energy functional** E_{tot} can be provided [61]

$$E_{\text{tot}} = \langle \hat{H} \rangle = \text{tr}(\hat{H}\hat{\varrho}) \tag{1.51}$$

$$= \text{tr}\left((\hat{T}^s + \hat{v}_{\text{H}} + \hat{v}_{\text{el}})\hat{\varrho}\right) + E_{\text{I}} + E_{\text{XC}}[\varrho]. \tag{1.52}$$

With the Laplacian[5] $\hat{L} = \frac{1}{2}\nabla^2$ the **kinetic energy** contribution becomes [61]

$$E_{\text{kin}} = \langle \hat{T}^s \rangle = \text{tr}(\hat{T}^s \hat{\varrho}) \tag{1.53}$$

$$= \sum_{i\sigma\mathbf{k}} \omega_{\mathbf{k}} f_{i\sigma\mathbf{k}}^{\text{occ}} \langle \Psi_{i\sigma\mathbf{k}} | \hat{L} | \Psi_{i\sigma\mathbf{k}} \rangle \tag{1.54}$$

$$= \sum_{i\sigma\mathbf{k}} \omega_{\mathbf{k}} f_{i\sigma\mathbf{k}}^{\text{occ}} |\mathbf{G} + \mathbf{k}|^2 |c_{i\sigma\mathbf{k}}|^2. \tag{1.55}$$

The **ion-ion contribution** to the total energy E_{I} can be decomposed into sums over $1/r$ potentials [1]

$$E_{\text{I}} = \frac{1}{2} \sum_{i \neq j} \frac{Z_i Z_j}{|\tau_i - \tau_j + \mathbf{R}|}. \tag{1.56}$$

The prefactor $\frac{1}{2}$ arises from the double counting of the ions in the above expression. In periodic systems such as crystals the series in the above term does not converge as it becomes an infinite sum over long-range $1/r$ potentials. Ewald [68, 69, 70] solved this problem by introducing an artificial screening charge with a Gaussian shape as the series of such screened $1/r$ potentials is converging. The Ewald summation is based on the following identity [61]:

$$\sum_i^\infty \frac{1}{|\tau_i - \tau_j + \mathbf{R}|} = \frac{2\pi}{\Omega} \sum_{\mathbf{G}} \int_0^\eta \exp(-\frac{|\mathbf{G}|^2}{4x^2}) \exp(i(\tau_i - \tau_j) \cdot \mathbf{G}) \frac{1}{x^3} dx$$

$$+ \frac{2}{\sqrt{\pi}} \sum_{\mathbf{R}} \int_\eta^\infty \exp(-|\tau_i - \tau_j + \mathbf{R}|^2 x^2) dx. \tag{1.57}$$

The non-converging infinite sum on the left-hand side is replaced by two infinite sums, the first one in the reciprocal space, the second one in real space. By choosing proper values of η the two sums on the right-hand side can converge rapidly in reciprocal or real space, respectively. This identity can be efficiently implemented in plane-wave codes, as various contributions to the Hamiltonian are evaluated in either real or reciprocal space. By introducing an artificial Gaussian screening charge ϱ_{Gauss} to the ionic contributions the following screened energy contributions can be obtained

$$E_{\text{I}} + E_{\text{H}}[\varrho] + E_{\text{el}}[\varrho] \stackrel{\text{Ewald}}{=} \tilde{E}_{\text{I}} + \tilde{E}_{\text{H}}[\varrho - \varrho_{\text{Gauss}}] + \tilde{E}_{\text{el}}[\varrho + \varrho_{\text{Gauss}}] - E_{\text{self}}. \tag{1.58}$$

---
[4] 1Ry=$\frac{1}{2}$Ha $\approx$ 13.6 eV.
[5] We follow the nomenclature of Ref. [67].

Entities with tilde are screened by the Gaussian. The last term E_{self} is due to the self-interaction between two Gaussians.

The energy contribution arising from the electron-electron interaction is obtained from the Poisson equation

$$\nabla v_{\text{H}}(\mathbf{R}) = -4\pi \tilde{\varrho}_{\text{H}}(\mathbf{R}) \tag{1.59}$$

with the screened charge density

$$\tilde{\varrho}_{\text{H}} = \varrho - \varrho_{\text{Gauss}}. \tag{1.60}$$

The Gaussian screening charge is constructed from spherical Gaussian with the radius r_{Gauss} and the charge z

$$\phi_{i_s}(r) = \frac{z_{i_s}}{\sqrt{\pi^3 r_{\text{Gauss},i_s}^3}} e^{-\frac{r^2}{r_{\text{Gauss},i_s}^2}}. \tag{1.61}$$

$$\varrho_{\text{Gauss}}(\mathbf{G}) = \sum_{i_s} \langle \mathbf{G} | \hat{T}_{i_s} | \phi_{i_s} \rangle \tag{1.62}$$

In order to project radial functions like $\phi(r)$ to the $\mathbf{G}$ space the following projector can be defined

$$\langle \mathbf{G} + \mathbf{k} | R_{nl} Y_{lm} \rangle = \sqrt{\frac{2l+1}{4\pi}} \frac{4\pi}{\sqrt{\Omega}} \int_0^\infty dr\, r^2 j_l(|\mathbf{G} + \mathbf{k}|r) R_{nl}(r) Y_{lm}(\theta_{\mathbf{G}}, \phi_{\mathbf{G}}). \tag{1.63}$$

with spherical Bessel functions j_l and spherical harmonics Y_{lm} [71].

The translator in Eq. (1.62) reads

$$\hat{T}_{i_s} = \sum_{i_a} e^{-i\mathbf{G} \cdot \tau_{i_s i_a}}. \tag{1.64}$$

The **Hartree potential** and energy become eventually

$$v_{\text{H}}(\mathbf{G} \neq \mathbf{0}) = \frac{4\pi}{|\mathbf{G}|^2} \tilde{\varrho}_{\text{H}}(\mathbf{G}). \tag{1.65}$$

$$E_{\text{H}} = \frac{1}{2} \text{tr}(\hat{v}_{\text{H}} \varrho). \tag{1.66}$$

The prefactor $\frac{1}{2}$ is due to the double counting correction.

The energy contributions arising from the exchange-correlation potential have been introduced already above (see Sec. 1.2.3). The remaining electron-ion contributions will be defined in the next section.

The choice of a plane-wave representation is, in particular, justified for 3d-periodic systems such as bulk crystals. Also in case of 2d periodic systems (e.g., surfaces) a plane-wave representation can be very efficient. Sometimes, even 1d-periodic systems, e.g., nanowires, can be efficiently treated using a plane-wave representation. However, one must not forget that the required vacuum regions (see Fig. 1.2(b)) are also sampled with plane waves. The usage of larger vacuum regions causes an increase of memory and computation demands.

Choosing plane waves as basis-set has various advantages:

- As already mentioned the completeness of the basis-set can be systematically controlled by one parameter, namely the energy cut-off E_{cut} and $\mathbf{k}$ point mesh.

- Plane-waves are orthogonal which simplifies the solution of the eigen problem. The issue of orthogonality will be discussed in the following chapter in detail.

- From the numerical point of view the contributions to the Hamiltonian are rather inexpensive to calculate. Particular the Hartree term can be elegantly expressed using the so-called Fourier derivative techniques [71] based on fast Fourier transformation (FFT). Such an approach scales only with $n_{pw} \ln n_{pw}$ when n_{pw} is the number of plane waves. Furthermore, the application of the Hamiltonian to the wave functions $|\xi\rangle = -\hat{H}|\Psi\rangle$ can be comfortably expressed using FFT.

- The basis-set does not directly depend on the atomic positions. Hence, when calculating forces no additional basis-set dependent terms occur.

However, plane waves have also some disadvantages. First of all, the computational effort depends on the system size. Larger systems require more memory and operations to be performed. Secondly, they are sampled on a uniform grid. An accurate description of core states would require a very high density of grid points in order to sample the nodal structure of the core states properly. In the following section it is described how a valence/core partitioning can help to benefit from the advantages of a plane-wave basis-set without introducing a too high sampling grid density.

1.5 Valence/core partitioning

According to the Bloch theorem a wave function can be expanded in terms of a *discrete* plane wave basis set when periodic boundary conditions are being applied. However, the high kinetic energy of the tightly bound electrons as well as the valence electrons in the core region lead to high frequencies of the corresponding wave functions. The resulting rapid oscillations would require an extremely and computationally infeasible large plane wave basis-set.

On the other hand chemical bonds of molecules and solids are to a greater extent determined by the valence electrons (with significant lower kinetic energy) rather than the core electrons. For the examination of many problems it, therefore, suffices to describe only these chemically active valence electrons quantum mechanically, while the chemically inert core electrons and the ions are being handled together with the nuclei as rigid non-polarized ion cores (pseudo potential). Such an approach is known as the frozen core approximation [72].

If, however, the core region is also required to be treated accurately and a pseudo potential approximation is not applicable all electrons have to be included. As an expansion of the tightly bound core orbitals in plane waves is computationally not feasible, the space is in this case partitioned into a muffin tin region describing the core wave functions and an interstitial region for the valence electrons. The muffin tin region is expanded, e.g., in terms of atomic orbitals sampled on a radial grid which can computationally efficiently sample rapid oscillations. The smoother wave functions in the interstitial region can be expanded in a relatively small plane wave basis-set. In addition projectors between the muffin tin and the interstitial region have to be defined. Methods based on such an approach are capable of describing wave functions of all orbitals and are hence referred to as all-electron methods.

Depending on the atomic system and the investigated observables as well as the required accuracy various methods to partition the valence and core region are available. Higher performance is usually traded with larger computational demands. The framework that will be developed within this work should be able to

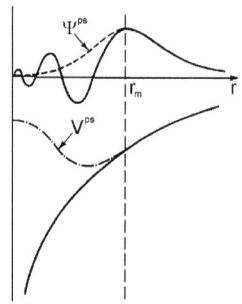
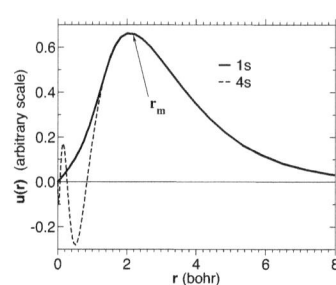

(a) Schematic illustration of the all-electron and pseudo electron potential

(b) Comparison of the 4s all electron and the pseudo 1s wave function of the Zn atom

Figure 1.3: (a) Schematic illustration of the all-electron (solid line) and pseudo electron potential (dashed line). The hard all-electron potential causes wave functions with depicting high frequency oscillations. By introducing a pseudo potential (V^{ps}) the obtained pseudo wave function Ψ^{ps} shows significantly lower frequencies which suggests an expansion into plane waves. The pseudo wave function Ψ^{ps} matches the all-electron wave function at the radius r_{m}. (b) Radial part of the all electron 4s (dashed line) and pseudo 1s wave function (solid line) as obtained from the Zinc pseudo potential generation.

provide different partitioning approaches for different tasks. Therefore, the most important approaches are discussed in this section. For many investigations related to semiconducting systems, such as those studied in this work, the influence of the core is rather small. Thus, this section first addresses the pseudo potential theory. The discussion will then focus on the description of the core region. Here, Slater's Augmented Plane Wave (APW) method will be sketched as well as the successive improvement steps over APW ranging from linearization (LAPW) to the application of full potentials (FP-LAPW), and eventually the introduction of local orbitals in the FP-LAPW+lo method. The discussion on valence/core partitioning will be completed with a short introduction to the very successful PAW method.

1.5.1 Pseudo-potential theory

Norm conserving pseudo-potentials. First-principles pseudo potentials are constructed on the basis of the scalar-relativistic radial Schrödinger equation of a single spherical atom. As a solution one obtains the all-electron potential as well as an all-electron wave function. The radial contributions to the wave functions of different magnetic quantum numbers m are identical due to the mentioned spherical symmetry.

A radial part of the pseudo wave function $u_l^{\mathrm{ps}}(r)$ is derived from the non-relativistic Schrödinger equation such that the following conditions are met:

1. **Eigenspectrum.** Both the pseudo wave function and the all-electron wave function yield the identical eigenvalue

$$\varepsilon_l^{\mathrm{ps}} \equiv \varepsilon_{nl}^{\mathrm{AE}}. \tag{1.67}$$

2. **Cut-off radius.** Outside the cut-off region (augmentation region) which is defined by cut-off radius

r^m, the pseudo wave function and the all-electron wave function match with respect to their amplitudes

$$u_l^{\mathrm{ps}}(\varepsilon_l^{\mathrm{ps}}, r) \to u_{nl}^{\mathrm{AE}}(r) \qquad \forall r > r_l^m \tag{1.68}$$

as well as their logarithmic derivatives

$$\frac{d}{dr}\ln u_l^{\mathrm{ps}} \to \frac{d}{dr}\ln u_{nl}^{\mathrm{AE}} \qquad \forall r > r_l^m. \tag{1.69}$$

By increasing the cut-off radius softer pseudo potentials can be generated which require a smaller plane wave basis-set. In return larger cut-off radii lead to a more inaccurate pseudo wave function in the region relevant to chemical bonding. Hence, the transferability of the pseudo potential suffers when increasing the cut-off radii.

3. **Norm conservation.** The pseudo and the all-electron wave functions are normalized

$$\int_0^\infty |u_l^{\mathrm{ps}}|^2 dr = \int_0^\infty |u_l^{\mathrm{AE}}|^2 dr = 1 \tag{1.70}$$

which implies norm conservation

$$\int_0^{r'} |u_l^{\mathrm{ps}}|^2 dr \equiv \int_0^{r'} |u_{nl}^{\mathrm{AE}}|^2 dr \qquad \forall r' > r_l^c. \tag{1.71}$$

4. **Nodal structure.** In contrast to the all-electron wave function the pseudo wave function has no radial nodes. It should be at least twice differentiable in order to make the pseudo potential be continuous. The pseudo valence states can be obtained from the u_l^{ps} via

$$|\Psi^{\mathrm{ps}}\rangle = \frac{1}{r}|u_l^{\mathrm{ps}} Y_{lm}\rangle \tag{1.72}$$

with Y_{lm} being the spherical harmonics [71].

5. **Pseudo potential contributions.** From the pseudo wave function the lth pseudo potential contribution v_l^{ps} is obtained by inverting the Schrödinger equation

$$\left(-\frac{1}{2}\nabla^2 + v_l^{\mathrm{ps}} - \varepsilon_l^{\mathrm{ps}}\right) u_l = 0 \tag{1.73}$$

$$\Longrightarrow v_l^{\mathrm{ps}} = \varepsilon_l^{\mathrm{ps}} + \frac{1}{2u_l^{\mathrm{ps}}}\nabla^2 u_l^{\mathrm{ps}}.$$

As for the pseudo wave function also the pseudo potential contributions must match the all-electron potential outside the cut-off region $r > r_l^m$.

For each valence state (s,p,d,f,..., $l_{\max}$) (at least) one pseudo potential v_l^{ps} is required. By applying a projector for each component the total pseudo potential can be decomposed into the single potential contributions

$$v^{\mathrm{ps}} = \sum_{l=0}^{l_{\max}} \sum_{m=-l}^{l} v_l^{\mathrm{ps}} |\chi_{lm}\rangle\langle\chi_{lm}|. \tag{1.74}$$

Semi local and fully separable form of the pseudo potential For $r \to \infty$ the shape of the potential can be nearly described as z^{val}/r. The long-range limit of the pseudo potential can be approximately treated independently of l. This component is called local pseudo potential $v_{\text{loc}}^{\text{ps}}$. Only the short-range limit remains l-dependent

$$v^{\text{ps}} = v_{\text{loc}}^{\text{ps}} + \sum_{l \neq l_{\text{loc}}}^{l_{\text{max}}} \sum_{m=-l}^{l} \Delta v_l^{\text{sl}} |\chi_{lm}\rangle\langle\chi_{lm}| \quad \text{with } \Delta v_l^{\text{sl}} = v^{\text{ps}} - v_{\text{loc}}^{\text{ps}}. \tag{1.75}$$

This decomposition is referred to as the semi local form of a pseudo potential. The potential is constructed such that Δv_l^{sl} vanishes beyond r_l^{m}. A further simplification can be achieved by restricting to the ground state only ($n=1$)

$$v^{\text{ps}} = v_{\text{loc}}^{\text{ps}} + \sum_{l}^{l_{\text{max}}} \sum_{m=-l}^{l} \Delta v_l^{\text{sl}} |\chi_{1lm}\rangle\langle\chi_{1lm}|. \tag{1.76}$$

In contrast to Eq. (1.74) the semi local form of the pseudo potential is computationally less demanding since the projectors belonging to $l = l_{\text{loc}}$ can be avoided. The number of projections to be performed is smallest if the maximum angular momentum is chosen as local component, i.e., $l_{\text{loc}} = l_{\text{max}}$. The required storage of the semi local form when being applied to a plane-wave basis-set is as large as $\frac{1}{2}(n_{\text{pw}}^2 + n_{\text{pw}})$.

An important simplification has been proposed by Kleinman and Bylander [36]. They suggested to treat also the radial pseudo potential as a non-local potential by replacing it with the projector

$$\langle \mathbf{R} | \Delta v_l^{\text{ps}} | \mathbf{R}' \rangle \Rightarrow \langle \mathbf{R} | \chi_l \rangle E_l^{\text{KB}} \langle \chi_l | \mathbf{R}' \rangle \tag{1.77}$$

and

$$\langle \mathbf{R} | \chi_l \rangle = \frac{1}{r} \frac{\Delta v_l^{\text{ps}} u_l^{\text{ps}}(\mathbf{R})}{\sqrt{\langle u_l^{\text{ps}} | \Delta v_l^{\text{ps}} | u_l^{\text{ps}} \rangle}} Y_{lm}. \tag{1.78}$$

In the Kleinman-Bylander form, also known as the fully separable form of the pseudo potential, the number of required projector evaluations reduces to $\sim n_{\text{pw}}$. On the other hand this truncation might yield wrong results because the order of atomic eigenstates is not necessarily correct. The Kleinman-Bylander Hamiltonian might obtain atomic eigenstates containing nodes which can lie energetically below the lowest state. Such *ghost states* are a direct consequence of the truncation of the pseudo potential. If such a ghost state is occupied within the self-consistent field calculation, a spurious density is obtained which leads to unphysical results. Thus, during the construction of the pseudo potentials it has to be ensured that such ghost states are not lying energetically below or near the physical valence states. This can be accomplished by a comparison of the atomic spectra for the semi local with the Kleinman-Bylander pseudo potential. Ghost states below the valence states can be identified with the criteria suggested by Gonze [73]. In most cases pseudo potentials free of ghost states can be generated by choosing a proper local component l_{loc} as well as cut-off radii.

Plane wave representation The framework which will be developed in the scope of this work will employ norm-conserving pseudo potentials. Since in this work the wave functions will be expanded in plane-waves, the local and non-local pseudo potential contributions have to be expressed in $\mathbf{G}$ space. The artificial Gaussian screening charge (see Eq. (1.62)) that has been introduced in the Hartree potential (Eq. (1.65)) via Eq. (1.60) can be conveniently subtracted in the **local pseudo potential** in order to maintain charge neutrality [30]

$$\phi_{\text{Gauss},i_s}(r) = \frac{z_{i_s}}{r}\text{erf}, \tag{1.79}$$

$$\langle r|\tilde{\phi}^{\text{ps}}_{\text{loc},i_s}\rangle = \tilde{\phi}^{\text{ps}}_{\text{loc},i_s}(r) = \sum_{i_s} \phi_{\text{loc},i_s}(r) + \phi_{\text{Gauss},i_s}(r), \tag{1.80}$$

$$v^{\text{ps}}_{\text{loc}}(\mathbf{G}) = \sum_{i_s} \sum_r \langle \mathbf{G}|\hat{T}_{i_s}|r\rangle \langle r|\tilde{\phi}^{\text{ps}}_{\text{loc},i_s}\rangle. \tag{1.81}$$

The energy contribution to the total energy E_{tot} is the corresponding expectation value

$$E_{\text{loc}} = \langle \hat{v}_{\text{loc}} \hat{\varrho} \rangle = \text{tr}(\hat{v}_{\text{loc}} \hat{\varrho}). \tag{1.82}$$

The **non-local pseudo potential** and energy contributions in a plane-wave basis are defined [30] as

$$\hat{v}_{\text{nl}} = \sum_{l \neq l_{\text{loc}}}^{l_{\text{max}}} \sum_{m=-l}^{l} \frac{|\Delta v^{\text{nl}}_{i_s l}|\Psi^{\text{ps}}_{i_s lm}\rangle \langle \Psi^{\text{ps}}_{i_s lm}|\Delta v^{\text{nl}}_{i_s lm}|}{\langle \Psi^{\text{ps}}_{i_s lm}|\Delta v^{\text{nl}}_{i_s l}|\Psi^{\text{ps}}_{i_s lm}\rangle} \tag{1.83}$$

and

$$E_{\text{nl}} = \langle \Psi|\hat{v}_{\text{nl}}|\Psi\rangle \tag{1.84}$$

$$= \sum_{i\sigma \mathbf{k}} \sum_{l \neq l_{\text{loc}}}^{l_{\text{max}}} \sum_{m=-l}^{l} \sum_{\mathbf{GG}'} \langle \Psi_{i\sigma \mathbf{k}}|\mathbf{G}+\mathbf{k}\rangle \frac{\langle \mathbf{G}+\mathbf{k}|\Delta v^{\text{nl}}_{i_s l}|\Psi^{\text{ps}}_{i_s lm}\rangle \langle \Psi^{\text{ps}}_{i_s lm}|\Delta v^{\text{nl}}_{i_s l}|\mathbf{G}'+\mathbf{k}\rangle}{\langle \Psi^{\text{ps}}_{i_s lm}|\Delta v^{\text{nl}}_{i_s l}|\Psi^{\text{ps}}_{i_s lm}\rangle} \langle \mathbf{G}'+\mathbf{k}|\Psi_{i\sigma \mathbf{k}}\rangle. \tag{1.85}$$

1.5.2 All-electron approaches

In the previous section the basic expressions that are necessary to implement a framework employing pseudo potentials in a plane-wave representation have been presented. In order to account for future implementations with respect to all-electron methods in the following the basic concepts of those methods are briefly sketched.

Since the pseudo potential approach does not treat the core states explicitly, properties which depend on them, such as hyperfine parameters, cannot be accurately expressed[6]. To overcome this problem also the core region has to be included into the quantum-mechanical description. In the following paragraphs various basis-sets to describe the core region are introduced. In particular, we sketch the often applied full-potential linearized augmented plane-wave method with local orbitals (FP-LAPW+lo). This approach is based on APW (Augmented Plane Waves) and has been successively been refined: By linearization of the energy dependence of the basis functions APW has been improved to the LAPW method. FP-LAPW describes in addition a full potential in the interstitial regions instead of a muffin-tin potential. The convergence of the semi-core states has been improved by additionally adding local orbitals in the FP-LAPW+lo method. In the following paragraphs we sketch the basic concepts of these methods. The Projector Augmented Wave (PAW) method that is able of improving the performance drastically with respect to LAPW will round up this discussion.

Slater's APW. In 1937 Slater introduced the augmented plane wave (APW) method [44]. The unit cell is partitioned in the interstitial region and augmentation spheres. In the interstitial region plane waves are taken as basis-set, while inside the augmentation sphere atomic partial waves of the form

---

[6]Approximations to hyperfine parameters can, however, be obtained within a pseudo potential approach (see for example [74]).

$$\phi^{\mathrm{APW}}(r) = \sum_{lm} A_{lm}^{\mathbf{k}} u_l(r,\varepsilon) Y_{lm} \tag{1.86}$$

are assumed. The $u_l(r,\varepsilon)$ are energy dependent radial basis functions. The free parameter $A_{lm}^{\mathbf{k}}$ makes sure that the plane waves $e^{i(\mathbf{G}+\mathbf{k})\cdot\mathbf{R}}$ match the atomic partial waves $u_l Y_{lm}$ at the augmentation sphere boundary. Inside the augmentation spheres the potential is assumed to reflect spherical symmetry, outside it is kept constant.

With $\mathbf{S}$ being the overlap matrix the following non-linear eigenvalue problem has to be solved

$$|\hat{H} - E\mathbf{S}| = 0. \tag{1.87}$$

The APW approach is computationally very demanding and numerically even unstable because the determinant's matrix is energy dependent. However, it was the starting point of a family of partitioning methods which are described below.

Linearization of the energy dependence. The energy dependence of the basis functions in the APW method leads to an expensive non-linear eigenvalue problem. To overcome this difficulty Anderson [75] suggested to map the APW set to an energy-independent basis set by linearizing the partial waves in energy. The energy-dependent partial wave $u_l^{\mathrm{APW}}(r,\varepsilon)$ can be expanded in a Taylor series about a reference energy ε_l

$$u(r,\varepsilon) = u_l(r,\varepsilon_l) + (\varepsilon - \varepsilon_l)\dot{u}_l(r,\varepsilon_l) + O((\varepsilon - \varepsilon_l)^2) \tag{1.88}$$

$$\dot{u}_l = \frac{\partial u_l}{\partial \varepsilon}. \tag{1.89}$$

In addition in APW the wave functions inside and outside the augmentation sphere are matched only with respect to the value but not with respect to the slope. This introduces additional contributions to the kinetic energy which have to be considered in the Hamiltonian. The linearization of the energy, i.e., Eq. (1.89), can be controlled by introducing the additional parameter $B_{lm}^{\mathbf{k}}$

$$\phi^{\mathrm{LAPW}}(r) = \sum_{lm}(A_{lm}^{\mathbf{k}} u_l(r,\varepsilon_l) + B_{lm}^{\mathbf{k}} \dot{u}_l(r,\varepsilon_l))Y_{lm}. \tag{1.90}$$

The parameters A and B are chosen such that the plane waves can be joined with respect to both value and slope. This procedure leads to a generalized eigenvalue problem

$$\hat{H}\mathbf{C} = E\mathbf{SC} \tag{1.91}$$

with $\mathbf{C}$ being the matrix of eigenvectors containing the wave function coefficient. The dimension of the involved matrices depends on the number of basis-set functions used to describe the interstitial region. Various functions have been applied as regular basis-sets (so-called envelope functions) ranging from plane-waves to Gaussian or Hankel functions. When applying plane-waves as envelope functions the method is referred to as LAPW [76], if Hankel functions are taken instead, the method is called LMTO (linear muffin-tin orbital) method [77]. Besides the number of envelope functions, the choice of the augmentation radius r_m controls the size of the matrices. Usually r_m is chosen such that it is about half the covalent radius.

The valence states are then described by the basis functions while the core states are localized inside the augmentation spheres. The linearization of the energy introduces additional constraints requiring more plane waves than in APW.

Full potential representation. So far, in both methods APW and LAPW/LMTO the potential has been treated in the so-called muffin tin approximation: Inside the augmentation region the potential is assumed to reflect spherical symmetry while within the interstitial region the true potential is approximated as a constant. The resulting shape of the potential suggests the name "muffin-tin" potential. A generalization of this approximation consists of expanding the potential in the interstitial region in terms of plane-waves like

$$V(\mathbf{R}) = \begin{cases} \sum_{lm} v_{lm}(|\mathbf{R}|) Y_{lm} & |\mathbf{R}| < r_m^\alpha \\ \sum_{\mathbf{k}} v_{\mathbf{k}} e^{i\mathbf{k}\cdot\mathbf{R}} & \text{interstitial region.} \end{cases} \quad (1.92)$$

The density can be represented analogously. The full description of the potential in the entire space gives this method the name Full Potential LAPW (FP-LAPW [78]).

Local Orbitals. A general drawback of LAPW is the treatment of semi-core states. These states lie energetically between the *delocalized* valence states and the core states *localized* inside the augmentation spheres. The semi-core states are not completely confined inside the spheres. These semi-core states have usually one principle quantum number below the valence state. In the case of Ti the 4p state is a valence state while the 3p state is a semi-core state.

In Ref. [79] Singh proposed the usage of local orbitals (lo) inside of the augmentation sphere. Local orbitals can treat two principle quantum numbers per l channel (in case of the Ti example, 3p and 4p). When such a semi-core state should be constructed the two corresponding reference energies ε_1 for the description of the 4p and ε_2 for the 3p state can be considered

$$\phi^{\text{LO}} = \sum_{lm} (\underbrace{A_{lm} u_l(r, \varepsilon_1) + B_{lm} \dot{u}_l(r, \varepsilon_1)}_{1^{\text{st}} \text{ ref. energy } \varepsilon_1} + \underbrace{C_{lm} u_l(r, \varepsilon_2)}_{2^{\text{nd}} \text{ ref. energy } \varepsilon_2}) Y_{lm}. \quad (1.93)$$

The free parameters A_{lm}, B_{lm}, and C_{lm} are constructed such that the local orbital has zero value and slope at the augmentation sphere radius r_m. Furthermore the local orbitals are strictly orthogonal which implies semi-core and valence states being orthogonal. The tail of semi-core states can be represented in the interstitial region using plane-waves.

FP-LAPW+lo. The previous ideas of APW, energy linearization, full-potential representation, as well as the usage of local orbitals to improve the semi-core convergence are merged in the FP-LAPW+lo method [45, 46]. Its basis-set is a mixture of plane waves in the interstitial region and a linear combination of APWs and local orbitals within the augmentation spheres

$$\phi(\mathbf{R}) = \begin{cases} \sum_{\mathbf{k}} c_{\mathbf{k}} e^{i\mathbf{k}\cdot\mathbf{R}} & \text{interstitial region} \\ \sum_{lm} \underbrace{(A_{lm} u_l + B_{lm} \dot{u}_l) Y_{lm}}_{\text{LAPW}} \underbrace{+ C_{lm}(A'_{lm} u_l + B'_{lm} \dot{u}_l + u_l(\varepsilon_2)) Y_{lm}}_{+\text{lo}} & |\mathbf{R}| < r_m. \end{cases} \quad (1.94)$$

The LAPW method is a powerful and very accurate all-electron scheme which has been used in a broad spectrum of applications. The high accuracy of LAPW is also the reason why it is used as benchmark method when comparing accuracies of other methods. The huge computational effort in the (FP-)LAPW(+lo) scheme arises from solving the generalized eigenvalue problem of Eq. (1.91) which has to be solved for huge matrices while obeying the matching constraints at r_m by adjusting the fitting parameters A_{lm}, B_{lm}, and C_{lm}.

PAW. The projector augmented waves (PAW) method proposed by Blöchl [31] generalizes both pseudo potentials and the above described augmentation methods. In PAW a transformation $\hat{T}$ between the true wave function $|\Psi\rangle$ and a numerically less demanding auxiliary wave function $|\tilde{\Psi}\rangle$ is introduced

$$|\Psi\rangle = \hat{T}|\tilde{\Psi}\rangle. \qquad (1.95)$$

The transformation should be chosen such that the smooth auxiliary wave function $|\tilde{\Psi}\rangle$ converges quickly with respect to the basis-set size. In order to yield the correct nodal structure of the true wave function, $\hat{T}$ has to modify $|\tilde{\Psi}\rangle$ inside each atomic region. Therefore, the atomic regions are described in terms of the *differences* between the true and the auxiliary wave functions, $|\phi\rangle$ and $|\tilde{\phi}\rangle$, respectively. Inside the atomic region the true wave function can be expanded in terms of the partial waves $|\phi\rangle$ which are the solutions of the radial Schrödinger equation for an isolated atom[7]

$$|\Psi\rangle = \sum_i c_i \langle r|\phi_i\rangle \qquad r < r_\mathrm{c}. \qquad (1.96)$$

For every partial wave $|\phi_i\rangle$ a smooth auxiliary partial wave $|\tilde{\phi}_i\rangle$ counterpart is constructed such that outside the atomic region both partial waves are identical and their difference cancels out identically

$$\left.\begin{array}{rcl}\phi_i(r) &=& \tilde{\phi}_i(r) \\ \Longrightarrow \quad \phi_i(r) - \tilde{\phi}_i(r) &=& 0\end{array}\right\} \quad r \geq r_\mathrm{c}. \qquad (1.97)$$

The above definition is a crucial element for the efficiency of PAW. In LAPW a cumbersome matching and fitting procedure is required in order to match the wave functions with respect to their value and slope. Otherwise the truncation of the wave functions at r_m would correspond to the introduction of an artificial multipole momentum. Throughout the PAW method always the *difference* of both partial waves is considered. Hence, a truncation error would be introduced in both partial waves simultaneously and hence, cancels out identically. As a result no fitting and matching is necessary in case of PAW.

Inside the atomic region the true wave function can be expressed in terms of the partial waves $|\phi\rangle$ and the auxiliary wave function $|\tilde{\Psi}\rangle$ can be expressed likewise in terms of the auxiliary partial waves $|\tilde{\phi}\rangle$

$$\left.\begin{array}{rcl}|\Psi\rangle &=& \sum_i |\phi_i\rangle c_i \\ |\tilde{\Psi}\rangle &=& \sum_i |\tilde{\phi}_i\rangle c_i\end{array}\right\} \quad r \geq r_\mathrm{c}. \qquad (1.98)$$

In order to project the true and the auxiliary wave functions from the interstitial to the atomic regions, projector functions $|\tilde{p}_i\rangle$ can be defined and the expansion coefficients c_i become

---

[7]Here the frozen core approximation is being applied. However, in principle the frozen core approximation can be relaxed in the PAW approach.

$$c_i = \langle \tilde{p}_i | \tilde{\Psi}_i \rangle \qquad \langle \tilde{p}_i | \tilde{\phi}_j \rangle = \delta_{ij}. \tag{1.99}$$

With the transformation defined in Eq. (1.95) the true wave function can now be expressed in terms of smooth auxiliary wave functions $|\tilde{\Psi}\rangle$, smooth auxiliary partial waves $|\tilde{\phi}_i\rangle$, smooth projector functions $|\tilde{p}_i\rangle$ as well as partial waves constructed as solutions of the radial Schrödinger equation for the isolated atom $|\phi\rangle$

$$|\Psi\rangle = |\tilde{\Psi}\rangle + \sum_i \left(|\phi_i\rangle - |\tilde{\phi}_i\rangle \right) \langle \tilde{p}_i | \tilde{\Psi} \rangle \tag{1.100}$$

$$= |\tilde{\Psi}\rangle + \sum_\tau \left(|\Psi^1_\tau\rangle - |\tilde{\Psi}^1_\tau\rangle \right). \tag{1.101}$$

Following Blöchl's notation one-center entities within the atomic region are labeled "1". The smooth auxiliary wave functions $|\tilde{\Psi}\rangle$ and the smooth partial waves $|\tilde{\phi}\rangle$ describe only the valence states. In order to evaluate expectation values also the n_c core states $|\phi^c\rangle$ have to be taken into account

$$\langle A \rangle = \underbrace{\sum_i \langle \tilde{\Psi}_i | \hat{T}^\dagger A \hat{T} | \tilde{\Psi}_i \rangle}_{\text{valence states}} + \underbrace{\sum_j^{n_c} \langle \phi^c_j | A | \phi^c_j \rangle}_{\text{core states}} \tag{1.102}$$

Analogously the remaining entities like the total energy and the charge density can be decomposed into the auxiliary contributions and one-center terms

$$E = \tilde{E} + \sum_\tau (E^1_\tau + \tilde{E}^1_\tau) \tag{1.103}$$

$$n(\mathbf{R}) = \tilde{n}(\mathbf{R}) + \sum_\tau (n^1_\tau(\mathbf{R}) + \tilde{n}^1_\tau(\mathbf{R})). \tag{1.104}$$

The actual expressions for the smooth auxiliary terms $\tilde{E}$ and $\tilde{n}$ as well as the one-center contributions E^1, $\tilde{E}^1$, n^1, and $\tilde{n}^1$ are derived in [31]. A complete review about the PAW method can be found in [80, 81].

In the PAW method all entities which have to be evaluated during the SCF cycles when diagonalizing the Kohn-Sham equations can be computed on a smaller basis-set of auxiliary functions (e.g., plane-wave basis-set with a energy cut-off of $\approx 30\,\text{Ry}$). The transformation operator $\hat{T}$ allows a direct access to the true wave function with the full nodal structure. In fact, it can be shown [80] that the total energy expression of the non-local pseudo potential can be obtained by expanding the PAW total energy expression into a Taylor series and truncating it beyond the linear term. From this point of view one might interpret PAW as a pseudo potential approach with a pseudo potential that adapts to the electronic environment at every SCF iteration. On the other hand PAW describes also an all-electron augmentation region like the APW family. PAW provides full access to the true wave function, full charge and spin densities as well as properties related to the core states, such as hyper fine parameters.

1.6 Tight-binding methods

With increasing number of atoms the application of *ab-initio* methods can become computationally too demanding. In the range up to 10^3 or even 10^7 atoms the tight-binding (TB) method [82] can be applied. Since the application of TB is important the framework should consider a future implementation of a corresponding TB Hamiltonian. Hence, we briefly sketch the major ideas behind TB in the following paragraphs.

Tight-binding can be interpreted as counterpart to the free-electron approximation. Its basic assumption is that the restricted Hilbert space that is spanned by atomic-like orbitals is sufficient to describe the solution of the Schrödinger equation within a restricted energy range. There are various tight-binding implementations, ranging from semi-empirical tight-binding to "*ab-initio*" based tight-binding. In semi-empirical tight-binding fitted parameters are used to describe matrix elements of the overlap and Hamilton operators directly. No localized basis is specified. Higher accuracy can be obtained by using a localized basis, such as atomic orbitals. This leads to the linear combination of atomic orbitals method (LCAO). In LCAO the Hamiltonian is expressed in terms of atomic orbitals μ and ν which yields the LCAO Hamilton matrix $\mathbf{H}^{\mathrm{LCAO}}$ as well as the overlap matrix $\mathbf{S}$

$$\mathbf{H}^{\mathrm{LCAO}}_{\mu\nu} = \langle\mu|\hat{H}|\nu\rangle \tag{1.105}$$

$$\mathbf{S}_{\mu\nu} = \langle\mu|\nu\rangle. \tag{1.106}$$

Basic assumption of this approach is that the overlap of orbitals is limited to only a few shells of neighboring atoms. If this assumption holds the tight binding Hamiltonian decomposes into a sparse matrix. For sparse matrices efficient eigensolvers exists which scale quadratically or linearly with the system size [83, 84, 85, 86].

Higher accuracy in the TB methods can be accomplished by introducing density functional theory into the TB method (DFTB). Therefore Foulkes and Haydock [87] have rewritten the expression of the total energy (Eq. (1.23)) by substituting the electronic charge density ϱ by a superposition of the reference densities ϱ^{ref} and small fluctuations $\delta\varrho^{\mathrm{ref}}$. E_{xc} is then expanded at this reference density up to the second order. Linear terms of ϱ^{ref} cancel out. The DFTB energy functional becomes then

$$\begin{aligned} E &= \sum_i \langle\Psi_i|\hat{H}[\varrho^{\mathrm{ref}}]|\Psi_i\rangle - \frac{1}{2}\int \frac{\varrho^{\mathrm{ref}}(\mathbf{R})\varrho^{\mathrm{ref}}(\mathbf{R}')}{|\mathbf{R}-\mathbf{R}'|}d\mathbf{R}d\mathbf{R}' + E_{\mathrm{xc}}[\varrho^{\mathrm{ref}}] \\ &- \int v_{\mathrm{xc}}[\varrho^{\mathrm{ref}}]\varrho^{\mathrm{ref}} + E_{\mathrm{ion-ion}} \\ &+ \frac{1}{2}\int \frac{1}{|\mathbf{R}-\mathbf{R}'|} + \left.\frac{\delta^2 E_{\mathrm{xc}}}{\delta\varrho\delta\varrho'^{\mathrm{ref}}}\right|_{\varrho^{\mathrm{ref}}} \delta\varrho\delta\varrho'^{\mathrm{ref}}. \end{aligned} \tag{1.107}$$

The linear terms in ϱ^{ref} cancel out. Traditional DFTB simply neglects the second order density dependent term (last line of latter equation) while in self-consistent charge tight-binding [5, 6, 7] (SCC-DFTB) the second order terms are considered in an extra charge density SCF loop.

1.7 Forces in ionic systems

So far we have focused on various methods to compute the electronic structure of atoms, molecules, and solids. These methods provide access to the system's energy as well as the wave function. Still missing is the access to the equilibrium geometry as well as a dynamic description of atomic positions. At a first glimpse a finite difference approach with respect to computed energies at varied atomic positions provides a straightforward access to forces which then can be used to integrate equations of motion (EOM). As pointed out already the self-consistent computation of the Born-Oppenheimer surface for different atomic positions can be computationally very demanding. Such an expensive approach can be avoided by exploiting perturbation theory.

In the following we discuss how forces can be obtained efficiently from first-principles calculations.

1.7.1 Hellmann-Feynman theorem

In classical mechanics forces acting on a particle at the coordinates τ can be obtained from the derivative of the potential energy U

$$\mathbf{F}_\tau = -\nabla_\tau U(\tau). \tag{1.108}$$

As analogon in quantum mechanics the forces can be determined according to

$$\mathbf{F} = -\nabla_\tau \langle E \rangle \quad \text{with } \langle E \rangle = \min \langle \Psi | \hat{H} | \Psi \rangle, \ \langle \Psi | \Psi \rangle = 1. \tag{1.109}$$

A proper ansatz to compute quantum mechanical forces is the Hellmann-Feynman theorem that was presented in 1937 by Feynman [88]. This theorem states that for any degree of freedom λ (in our case the atomic coordinates τ) the following identity holds

$$\frac{\partial E}{\partial \lambda} = \langle \frac{\partial \Psi}{\partial \lambda} | \overbrace{\hat{H} | \Psi \rangle}^{= E|\Psi\rangle} + \langle \Psi | \frac{\partial \hat{H}}{\partial \lambda} | \Psi \rangle + \overbrace{\langle \Psi | \hat{H}}^{= E\langle \Psi|} | \frac{\partial \Psi}{\partial \lambda} \rangle \tag{1.110}$$

$$= \langle \Psi | \frac{\partial \hat{H}}{\partial \lambda} | \Psi \rangle + E \frac{\partial}{\partial \lambda} \langle \Psi | \Psi \rangle. \tag{1.111}$$

Hence, the Hellmann-Feynman theorem becomes eventually

$$\frac{\partial E}{\partial \lambda} = \langle \Psi | \frac{\partial \hat{H}}{\partial \lambda} | \Psi \rangle. \tag{1.112}$$

Applied to atomic coordinates the Hellmann-Feynman theorem implies that the forces can be computed directly from the ground state wave functions, which are available from the total energy calculations anyway.

Finite basis-sets The Hellmann-Feynman theorem is only valid if Ψ is an *exact* eigenstate. For variational calculations of the ground state energy E, Ψ is expanded in a finite basis-set. In this case Eq. (1.110) cannot be simplified to Eq. (1.112) anymore since the first and last term of Eq. (1.110) have to be considered explicitly. Here the term

$$\langle \frac{\partial \Psi}{\partial \lambda} | \hat{H} | \Psi \rangle - \langle \Psi | \hat{H} | \frac{\partial \Psi}{\partial \lambda} \rangle \tag{1.113}$$

becomes a matrix expression and does not vanish. There are two approaches when computing forces from ground state energy calculations. (1) The basis-set can be constructed such that Ψ does not depend on λ. Such a requirement is known as the Hurley condition [89, 90]. The plane-wave basis-set fulfills this condition and the forces can obtained from the Hellmann-Feynman forces $\mathbf{F}^{\text{HF}}$

$$\mathbf{F} = \mathbf{F}^{\text{HF}} = -\frac{\partial E}{\partial \tau} = -\langle \Psi | \frac{\partial \hat{H}}{\partial \tau} | \Psi \rangle. \tag{1.114}$$

(2) In case of atomic-centered basis functions the Hellmann-Feynman forces of Eq. (1.112) will not provide correct forces. Here the full Eq. (1.110) has to be considered. The term (1.113) is called as the Pulay force $\mathbf{F}^{\text{Pulay}}$

$$\mathbf{F} = \mathbf{F}^{\text{HF}} + \mathbf{F}^{\text{Pulay}} = -\langle \Psi | \frac{\partial \hat{H}}{\partial \tau} | \Psi \rangle + \left(\langle \frac{\partial \Psi}{\partial \tau} | \hat{H} | \Psi \rangle - \langle \Psi | \hat{H} | \frac{\partial \Psi}{\partial \tau} \rangle \right). \tag{1.115}$$

1.8 Conclusions

In this chapter a brief summary of theories has been presented that are necessary to derive a flexible framework for developing efficient CMD applications. Depending on the required accuracy, in electronic structure calculations the description of the valence/core partitioning is crucial. Various methods ranging from pseudorization of the core to a full all-electron description have been sketched. Since the library should be able to cover a wide range of system sizes, we also presented an overview of important approximations, such as the tight-binding approach. Based on the here mentioned methods in the next chapter common issues of these methods will be identified and generalized to provide an more flexible approach to electronic structure simulations.

In the scope of this work we limit ourselves to the pseudo potential method while keeping the other methods in mind. It will be shown that with our approach other basis-sets and/or potentials can be easily be covered.

Chapter 2

Methods

In order to obtain physical properties based on the theoretical concepts described in the previous chapter, numerical/physical methods have to be applied which will be introduced in this chapter. We provide a brief overview on the methods which allow the development a general framework for CMD applications. The following discussions focus on methods to

- provide access to the electronic structure of a material (Sec. 2.1),
- describe structural properties of a system (Sec. 2.2), and to
- obtain thermodynamic properties of materials (Sec. 2.3).

2.1 Electronic minimization schemes

We begin the discussion on methods with an overview on how ground state properties E_{tot}, ϱ, and $\varepsilon_{i\sigma}(\mathbf{k})$ can be computed efficiently. With these entities a broad spectrum of material properties at $T = 0\,\text{K}$ can be derived, such as equilibrium lattice parameters, bulk moduli, cohesive energies, and band gaps.

If the system is described within DFT, a direct approach to compute the ground state wave functions belonging to the minimum of the total energy is to diagonalize the matrix $\mathbf{H}_{\mathbf{GG}'} = \langle \mathbf{G}|\hat{H}|\mathbf{G}'\rangle$. However, one has to keep in mind that calculations of even small systems require a rather large basis-set ($10^4 - 10^5$ plane waves). As the matrix $\langle \mathbf{G}|\hat{H}|\mathbf{G}'\rangle$ is not sparse its memory demand scales like $O(N_{\text{pw}}^2)$. Also the computational effort which is necessary to diagonalize a general matrix scales like $O(N_{\text{pw}}^3)$. An alternative way of finding the ground state is the usage of iterative schemes (see Fig. 2.1).

The residual vector R in

$$0 \stackrel{!}{=} R = \hat{H}|\Psi^{\text{trial}}\rangle - \varepsilon|\Psi^{\text{trial}}\rangle$$

vanishes if the trial wave function $|\Psi^{\text{trial}}\rangle$ is identical to the ground state wave function $|\Psi\rangle$. The negative[1] gradient of the energy with respect to the wave functions can be evaluated using the variational principle for normalized Kohn-Sham wave functions

$$|\xi_{i\sigma\mathbf{k}}\rangle = -|g_{i\sigma\mathbf{k}}\rangle = \frac{\delta \varepsilon_{i\sigma\mathbf{k}}}{\delta \langle \Psi_{i\sigma\mathbf{k}}|} = \frac{\delta}{\delta \langle \Psi_{i\sigma\mathbf{k}}|} \langle \Psi_{i\sigma\mathbf{k}}|\hat{H}|\Psi_{i\sigma\mathbf{k}}\rangle = \hat{H}|\Psi_{i\sigma\mathbf{k}}\rangle. \qquad (2.1)$$

---
[1] by definition the gradient always points upwards

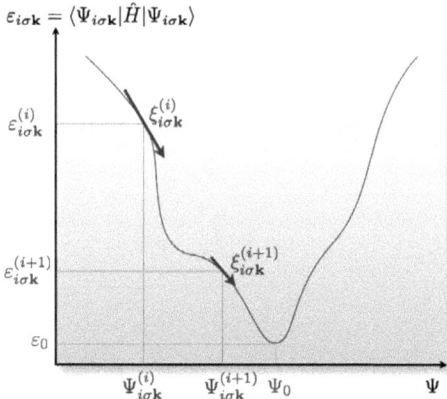

Figure 2.1: Schematic representation of the one-particle energy as function of the wave functions. Each wave function is a vector in a multidimensional vector space. An initial wave function $\Psi^{(1)}$ is improved by using the negative gradient, denoted with $|\xi_{i\sigma\mathbf{k}}\rangle$ until the minimum is found.

In all iterative minimization schemes the negative gradient is used to improve the wave function of the nth iteration

$$|\Psi_{i\sigma\mathbf{k}}^{(n+1)}\rangle = |\Psi_{i\sigma\mathbf{k}}^{(n)}\rangle - M(|\xi_{i\sigma\mathbf{k}}^{(n)}\rangle). \tag{2.2}$$

The iterative minimizers M differ basically in the way how the negative gradient is being used and whether a single state/band is updated sequentially (state-by-state schemes) or all states/bands are improved at once (all-state schemes).

In the following paragraphs important aspects of multi-dimensional minimization techniques are summarized which will be applied in this work, i.e., the numerically efficient evaluation of gradients (Sec. 2.1.1), the set up of search direction vectors (Sec. 2.1.2), the introduction of preconditioning (Sec. 2.1.3), and conjugating subsequent search vectors (Sec. 2.1.4).

2.1.1 Gradients $-\frac{\delta}{\delta\langle\Psi|}$

In order to support iterative minimization algorithms the gradient of $\hat{H}$ needs to be evaluated.

The gradient of the Hamiltonian can be obtained from

$$\frac{\delta\langle\Psi_{i\sigma\mathbf{k}}|\hat{H}|\Psi_{i\sigma\mathbf{k}}\rangle}{\delta\langle\Psi_{i\sigma\mathbf{k}}|} = -\hat{H}|\Psi_{i\sigma\mathbf{k}}\rangle. \tag{2.3}$$

It will be shown in the next chapter that the matrix form of the gradient has significant advantages for the runtime performance of the framework. The gradient in matrix form reads

$$(\xi_{i\mathbf{G}})_{\sigma\mathbf{k}} = -(\hat{H}_{\mathbf{GG'}}|\Psi_{i\mathbf{G}})_{\sigma\mathbf{k}}. \tag{2.4}$$

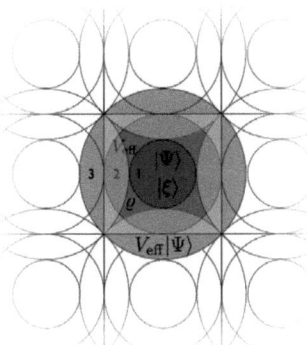

Figure 2.2: Illustration of wrap-around errors by a two-dimensional sketch of the (periodic) Fourier space. Wave functions $|\Psi\rangle$ and gradients $|\xi\rangle$ are sampled within a cut-off sphere with the radius $1G_{\text{cut}}$ (innermost circle 1). Components of the charge density $\varrho(\mathbf{G})$ and the effective potential $v_{\text{eff}}(\mathbf{G})$ are defined inside a cut-off sphere of twice the size (circle 2). In order to sample the gradient $v_{\text{eff}}(\mathbf{G})\Psi(\mathbf{G})$ a sphere of $3G_{\text{cut}}$ would be necessary (circle 3).
Using a FFT grid with only $2G_{\text{cut}}$ the high frequency contributions between $2G_{\text{cut}}$ and $3G_{\text{cut}}$ would be folded back into the first cell (yellow area) due to the periodicity of the Fourier space. Hence, an artificial wrap-around error within the interval $2G_{\text{cut}} < G_{\text{wrap}} < 3G_{\text{cut}}$ would occur. Because the gradient is sampled only up to $1G_{\text{cut}}$ the back-folded components between $2G_{\text{cut}}$ and $3G_{\text{cut}}$ do not contribute. Thus, instead of using a (more expensive) FFT grid of $3G_{\text{cut}}$ a smaller one of $2G_{\text{cut}}$ can be applied.

In the plane-wave representation the effective potential is diagonal in real space, i.e., the matrix $\langle\mathbf{R}|\hat{v}_{\text{eff}}|\mathbf{R}'\rangle = v_{\text{eff}}(\mathbf{R})\delta(\mathbf{R} - \mathbf{R}')$, whereas the kinetic gets diagonal in reciprocal space (see p. 20). Hence, the computation of the gradient can be efficiently evaluated using Fourier transformations [30]

$$\langle\mathbf{G} + \mathbf{k}|\xi_{i\sigma\mathbf{k}}^{\text{kin}}\rangle = \langle\mathbf{G} + \mathbf{k}|\hat{L}|\Psi_{i\sigma\mathbf{k}}\rangle, \quad (2.5)$$

$$\langle\mathbf{G} + \mathbf{k}|\xi_{i\sigma\mathbf{k}}^{\text{nl}}\rangle = \langle\mathbf{G} + \mathbf{k}|\hat{v}_{\text{nl}}|\Psi_{i\sigma\mathbf{k}}\rangle, \quad (2.6)$$

$$\langle\mathbf{R}|\xi_{i\sigma\mathbf{k}}^{\text{eff}}\rangle = \langle\mathbf{R}|\hat{v}_{\text{eff},\sigma}|\Psi_{i\sigma\mathbf{k}}\rangle. \quad (2.7)$$

Convolution problem $v_{\text{eff}}(\mathbf{G})$ is defined with Fourier coefficients up to $2G_{\text{cut}}$, $\Psi(\mathbf{G})$ up to G_{cut}. The gradient $\xi^{\text{eff}}(\mathbf{G})$ that is to be used to improve the wave function, is also sampled up to G_{cut}. According to the Fourier folding rule the highest frequency components of

$$|\xi\rangle = \hat{v}_{\text{eff}}|\Psi\rangle \quad (2.8)$$

range to $3G_{\text{cut}}$. Fortunately, it is possible to restrict this evaluation to a smaller (and faster) $2G_{\text{cut}}$ FFT grid.

When a smaller $2G_{\text{cut}}$ FFT grid is used, a wrap around error (see Fig. 2.2) would occur. However, in the gradient $|\xi\rangle$ only components of max. $1G_{\text{cut}}$ are taken into account. The wrap-around error contributions exist only in the frequency range [91] of

$$2G_{\text{cut}} < G_{\text{wrap}} < 3G_{\text{cut}}. \qquad (2.9)$$

Hence, the error contributions cannot affect the gradient $|\xi\rangle$ and the smaller FFT grid of $2G_{\text{cut}}$ suffices! The consideration of the convolution problem is important with respect to the computational efficiency of implemented gradients.

2.1.2 Search direction

In the case of the steepest descent scheme [92] the wave functions are iterated by "walking down" along the negative residual vector direction with a fixed step width. Hence, the search direction vector $|X\rangle$ is

$$\begin{aligned}|X_{i\sigma\mathbf{k}}^{(n)}\rangle &= (\hat{H} - \varepsilon_{i\sigma\mathbf{k}}^{(n)})|\Psi_{n\sigma\mathbf{k}}^{(n)}\rangle \\ &= |\xi_{i\sigma\mathbf{k}}\rangle - \varepsilon_{i\sigma\mathbf{k}}^{(n)}|\Psi_{i\sigma\mathbf{k}}^{(n)}\rangle. \end{aligned} \qquad (2.10)$$

The $\varepsilon_{i\sigma\mathbf{k}}^{(n)}$ denote the approximation of the one-particle energies in the nth iteration

$$\varepsilon_{i\sigma\mathbf{k}}^{(n)} = \langle \Psi_{i\sigma\mathbf{k}}^{(n)}|\hat{H}|\Psi_{i\sigma\mathbf{k}}^{(n)}\rangle. \qquad (2.11)$$

In the steepest descent scheme the wave functions are improved as

$$|\Psi_{i\sigma\mathbf{k}}^{(n+1)}\rangle = |\Psi_{i\sigma\mathbf{k}}^{(n)}\rangle - \delta t|X_{i\sigma\mathbf{k}}^{(n)}\rangle. \qquad (2.12)$$

In the latter equation δt is the artificial time step. The larger δt, the larger is the change in the wave functions per iteration step $|\Psi^{(n+1)}\rangle - |\Psi^{(n)}\rangle$. However, if δt exceeds a critical value the scheme becomes unstable. The position of that critical value depends strongly on the system.

2.1.3 Preconditioning

For realistic systems the steepest descent scheme converges too slowly [92]. In order to improve it, a basic understanding of the plane-wave Hamiltonian's shape is imperative. As already pointed out in Eq. (1.46) the kinetic energy contribution $\hat{T}_{\mathbf{GG}'}$ contains only the diagonal elements

$$\hat{T}_{\mathbf{GG}'} = \delta_{\mathbf{GG}'}|\mathbf{G} + \mathbf{k}|^2 c_{i\sigma\mathbf{k}}(\mathbf{G}).$$

The plane-wave basis $\langle\mathbf{G} + \mathbf{k}|$ is expanded up to the energy cut-off E_{cut}. According to Eq. (1.49) higher energy cut-offs introduce a $\langle\mathbf{G} + \mathbf{k}|$ basis with large $|\mathbf{G} + \mathbf{k}|^2$ values. In case of high energy cut-offs this leads to a domination of the kinetic energy contributions to $\hat{H}_{\mathbf{GG}'}$ over those of the effective potential $v_{\text{eff}}(\mathbf{G} - \mathbf{G}')$ in the high frequency regime of the plane-wave Hamiltonian matrix. In other words, high energy cut-offs cause a domination of the diagonal over the off-diagonal elements. Such a matrix becomes ill-conditioned[2] and is difficult to diagonalize. The actual problem can be easily demonstrated in the 2-dimensional case. For this purpose the matrix $\langle\mathbf{G}|\hat{H}|\mathbf{G}'\rangle$ with the dominating diagonal shall be iconified with a two–dimensional

---
[2]The condition of a matrix is the ratio of its largest and its smallest eigenvalue.

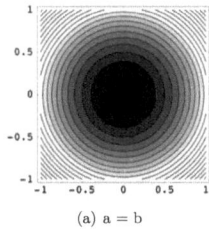 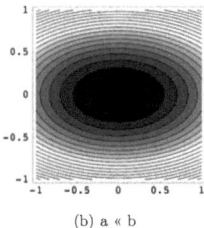

(a) a = b   (b) a « b

Figure 2.3: Illustration of the influence of the matrix conditioning to the convergence rate.

diagonal matrix

$$\langle \mathbf{G}|\hat{H}|\mathbf{G'}\rangle \rightarrow \begin{pmatrix} a & 0 \\ 0 & b \end{pmatrix}$$

and the basis-set functions are represented with a simple 2d vector

$$\varepsilon = \langle \Psi|\mathbf{G}\rangle\langle \mathbf{G}|\hat{H}|\mathbf{G'}\rangle\langle \mathbf{G'}|\Psi\rangle \rightarrow \varepsilon = (x,y) \begin{pmatrix} a & 0 \\ 0 & b \end{pmatrix} \begin{pmatrix} x \\ y \end{pmatrix}.$$

Evaluating the latter expression yields

$$\varepsilon = ax^2 + by^2.$$

If a and b are identical the solution is a sphere whereas the cases $a \ll b$ or $a \gg b$ (ill-conditioning of the matrix) form an ellipsoid. Following the negative gradient direction on an ellipsoidal surface leads to a slow "zick-zack-like" minimization path involving many iterations (Fig. 2.4). That leads to the conclusion that the steepest-descent scheme can/should be applied only when the energy cut-off can be chosen very small (≤ 5 Ry). Yet, for realistic systems much higher energy cut-offs are required. To cope with the bad matrix conditions a preconditioner K can be applied. A preconditioner approximates $\hat{H}_{\mathbf{GG'}}^{-1}$. In the case of a plane-wave representation it becomes a function of the kinetic energy. A preconditioner can be seen as a mapping function to reshape the ellipsoidal form back to a spherical one, or equally, to decrease the ratio of the largest and smallest eigenvalue of the Hamiltonian matrix in order to improve its condition. Williams–Soler [93] combined the steepest descent with the preconditioner

$$K(\mathbf{G}) = \frac{1 - e^{-\alpha(\mathbf{G})\lambda}}{\alpha(\mathbf{G})} \qquad (2.13)$$

$$\alpha(\mathbf{G}) = \mathbf{H}_{\mathbf{GG}} - \varepsilon_{i\sigma\mathbf{k}}^{(i)}. \qquad (2.14)$$

The parameter λ is a parameter indicating the step length. Larger values converge faster but might light to numerical instabilities. Typical values of λ are between 0.1 and 10. The search direction in the Williams–Soler scheme reads then

$$|X_{i\sigma\mathbf{k}}^{(n)}\rangle = K|\xi_{i\sigma\mathbf{k}}^{(n)}\rangle + \varepsilon_{i\sigma\mathbf{k}}^{(n)}|\Psi_{i\sigma\mathbf{k}}^{(n)}\rangle. \qquad (2.15)$$

The Williams–Soler algorithm is more stable than the steepest-descent and can be used for higher energy cut-offs. It has the same memory demand as the steepest descent scheme.

The concept of preconditioning is crucial for many minimization schemes. In the following sections various

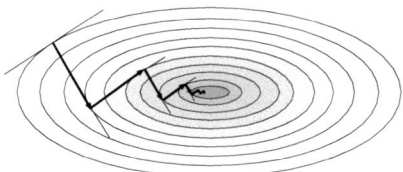

Figure 2.4: Sketch of a zick-zack minimization path of a steepest descent scheme with optimized step length. In the optimized steepest-descent the artificial time step δt is chosen by a line minimization. Therefore, the search direction of the next iteration is perpendicular to the previous one. The more ellipsoidal the surfaces gets the stronger is the "zick-zack-like" of the minimization paths which leads to a slower convergence rate.

more advanced preconditioners will be introduced.

2.1.4 Conjugate gradient methods

The previously described minimization schemes suffer from the fact that each minimization step along the search direction might affect degrees of freedom which were optimized already in a previous step. Hence, a subsequent minimization step can *re*introduce new errors proportional to the previous search direction. To avoid this coupling between subsequent search directions two conditions have to meet: along each vector the line minimum has to be found and the new search direction must be chosen *conjugate* rather than perpendicular to the previous one. Conjugate means that coefficients, whose residual vector contributions were already negligible, will not be considered in further iterations anymore. Hence, the multidimensional vector space itself in which the minimization takes place is reduced. Theoretically, an optimal (exponential[3]) convergence rate can so be accomplished. That is the idea of all conjugate gradient methods [94, 92].

Assuming the function to be minimized can be approximated by a multidimensional quadratic function f at the point P

$$f(x) = c - \langle b|X\rangle + \frac{1}{2}\langle X|A|X\rangle. \tag{2.16}$$

$$c = f(P),\ b = -\left.\nabla f\right|_P,\ A_{ij} = \left.\frac{\partial^2 f}{\partial x_i \partial x_j}\right|_P,\ X = (x_1, x_2, \ldots, x_m).$$

A refers to the $m \times m$ Hessian matrix. Within an iteration the improved point $P^{(n)}$ is obtained from the current value $P^{(n-1)}$ via

$$P^{(n)} = P^{(n-1)} + \lambda^{(n-1)} h^{(n-1)} \tag{2.17}$$

The iteration step is denoted as n, $\lambda^{(n)}$ is a step width and $h^{(n)}$ is the vector of the search direction. The search direction is obtained from

$$h^{(n)} = \begin{cases} -\nabla f(P^{(n)}) & n = 0 \\ -\nabla f(P^{(n)}) + \gamma^{(n-1)} h^{(n-1)} & n > 0, \end{cases}$$

i.e., for the first iteration a steepest descent vector acts as search direction. For all subsequent iterations the

---

[3] i.e., $\ln(E^{(n)} - E^{\text{converged}})$ converges linearly.

search direction vector contains an additional constraint $\gamma^{(n)}h^{(n)}$ with

$$\gamma^{(n)} = \frac{\langle g^{(n)}|g^{(n)}\rangle}{\langle g^{(n-1)}|\gamma^{(n-1)}\rangle} \tag{2.18}$$

$$g^{(n)} = -\nabla f(P^{(n)}). \tag{2.19}$$

This additional constraint ensures the new search direction $h^{(n)}$ being *conjugate* to all previous ones, i.e.,

$$\langle h^{(l)}|A|h^{(n)}\rangle = 0 \quad \forall l \neq n. \tag{2.20}$$

All-state conjugate gradient

It will be shown later (Sec. 3.1.3) that the computational efficiency of evaluating complex algebraic equations can be significantly improved using (blocked) matrix operations. Therefore, the minimization problem can be reformulated to a matrix notation. Treating all states i simultaneously allows to rewrite the object $\langle \mathbf{G} + \mathbf{k}|\Psi_{i\sigma\mathbf{k}}\rangle$ as a matrix

$$\langle \mathbf{G} + \mathbf{k}|\Psi_{i\sigma\mathbf{k}}\rangle \Rightarrow \mathbf{C}_{\mathbf{G}i}(\sigma\mathbf{k}). \tag{2.21}$$

In case of plane-waves the coefficient matrix $\mathbf{C}$ has the dimensions $n_{\text{pw}} \times n_{\text{occ-states}}$. Also the other ingredients to a preconditioned conjugate gradient method can be rewritten in form of matrices

$$|g_{i\sigma\mathbf{k}}\rangle = \hat{H}|\Psi_{i\sigma\mathbf{k}}\rangle \quad \Rightarrow \quad \mathbf{g}_{\mathbf{G}i} = \hat{H}\mathbf{C}_{\mathbf{G}i} \tag{2.22}$$

$$|X_{i\sigma\mathbf{k}}\rangle \quad \Rightarrow \quad \mathbf{X}_{\mathbf{G}i}. \tag{2.23}$$

The all-state search direction in matrix notation reads then [67]

$$\mathbf{X}_{\mathbf{G}i} = K\mathbf{g}_{\mathbf{G}i} + \frac{\operatorname{Re}\operatorname{tr}\left((\mathbf{P}_{\mathbf{G}i} - \mathbf{g}_{\mathbf{G}i})^\dagger \mathbf{g}_{\mathbf{G}i}\right)}{\operatorname{Re}\operatorname{tr}\left((\mathbf{P}_{\mathbf{G}i}^{\text{old}})^\dagger \mathbf{g}_{\mathbf{G}i}^{\text{old}}\right)} \mathbf{X}^{(n-1)}. \tag{2.24}$$

The wave function coefficient matrix is updated by

$$\mathbf{C}_{\mathbf{G}i}^{\text{new}} = \mathbf{C}_{\mathbf{G}i} + \lambda \mathbf{X}_{\mathbf{G}i} \tag{2.25}$$

with λ is chosen to yield the line minimum along the search direction $\mathbf{X}$ in order to conjugate the vector of the next iteration [92]. It will be shown later (Sec. 3.3.1) that for most applications a quadratic line minimization is sufficient[4]. For the quadratic fit as used in Eq. (41) the total energy value and its derivative as well as a distant trial energy E_{trial} are considered. The derivative reads

$$\mathbf{D} = 2\operatorname{Re}\operatorname{tr}(\mathbf{X}^\dagger \mathbf{G}) \tag{2.26}$$

whereas the trial energy is

$$E_{\text{trial}} = E_{\text{tot}}[(\mathbf{C} - \lambda_{\text{trial}}\mathbf{X})^\perp]. \tag{2.27}$$

---
[4]The application of a quadratic line minimization is also suggested in Ref. [67].

The symbol $\overline{\mathbf{X}}^\perp$ refers to the required orthonormalization of the wave function $\overline{\mathbf{X}}$. The minimum value of the quadratic fit is then

$$\lambda = \frac{\mathbf{D}}{2c} \qquad (2.28)$$

with the curvature

$$c = \frac{1}{\lambda_{\text{trial}}}(E_{\text{trial}} - (E + \lambda_{\text{trial}}\mathbf{D})). \qquad (2.29)$$

The electronic charge density is updated according to

$$\varrho_\sigma(\mathbf{R}) = \sum_{i\mathbf{k}} \omega_\mathbf{k} |\Psi_{i\sigma\mathbf{k}}(\mathbf{R})|^2. \qquad (2.30)$$

State-by-state conjugate gradient

Beside the (memory consuming) all-band conjugate gradient a state-by-state conjugate gradient scheme [26] that can treat systems with partially occupied or even empty states will be implemented in this work. In contrast to the previously described minimization schemes the Hamiltonian is diagonalized at a constant charge density $\hat{H}(\varrho_{\text{in}} = \text{const})$. Hence, for a fixed Hamiltonian the total energy $E_{\text{tot}}[\varrho_{\text{in}} = \text{const}]$ cannot be the variable to be minimized anymore. Instead, the one-particle energies $\varepsilon_{i\sigma\mathbf{k}}$ can be considered.

Since the matrix elements $\langle \Psi_i^{(n)} | \hat{H}[\varrho_{\text{in}}] | \Psi_j^{(n)} \rangle$ are not the correct one-particle energies, a subspace rotation has to be performed. In order to decouple the $|\Psi^{(n)}\rangle$ and to get access to the correct ε, a uniform rotation matrix $\mathbf{U}$ is constructed [67] in the subspace spanned by the set of wave functions of the current iteration $|\Psi^{(n)}\rangle$

$$|\xi^{(n)}\rangle = \hat{H}|\Psi^{(n)}\rangle, \qquad (2.31)$$
$$U_{ij} = \langle \xi_i | \xi_j \rangle. \qquad (2.32)$$

The decoupled wave functions $|\Psi'\rangle$ which yield to correct $\varepsilon = \langle \Psi' | \hat{H}[\varrho_{\text{in}}] | \Psi' \rangle$ can be obtained by rotating along the eigenvectors of $\mathbf{U}$.

$$\mathbf{U}\mathbf{u} = u\mathbf{u} \qquad (2.33)$$
$$|\Psi'\rangle = \mathbf{u}|\Psi\rangle. \qquad (2.34)$$

As before, a preconditioned conjugate gradient scheme is applied, just only for a single state

$$|g_{i\sigma\mathbf{k}}^{(n)}\rangle = \hat{H}[\varrho_{\text{in}}]|\Psi'_{i\sigma\mathbf{k}}\rangle \qquad (2.35)$$
$$|P_{i\sigma\mathbf{k}}^{(n)}\rangle = K|g_{i\sigma\mathbf{k}}^{(n)}\rangle \qquad (2.36)$$
$$|X_{i\sigma\mathbf{k}}^{(n)}\rangle = K|g_{i\sigma\mathbf{k}}^{(n)}\rangle + \frac{\langle P_{i\sigma\mathbf{k}}^{(n)} - g_{i\sigma\mathbf{k}}^{(n)} | g_{i\sigma\mathbf{k}}^{(n)} \rangle}{\langle P_{i\sigma\mathbf{k}}^{(n-1)} | g_{i\sigma\mathbf{k}}^{(n-1)} \rangle} |X_{i\sigma\mathbf{k}}^{(n-1)}\rangle. \qquad (2.37)$$

The improved wave function $|\Psi^{(n)}\rangle$ can be obtained from a linear combination of the previous wave function $|\Psi^{(n-1)}\rangle$ and the conjugate search direction $|X^{(n)}\rangle$ [92]

$$|\Psi^{(n)}\rangle = \alpha|\Psi^{(n-1)}\rangle + \beta|X^{(n)}\rangle. \qquad (2.38)$$

Since $\hat{H}|\Psi\rangle$ can be evaluated efficiently[5] for a fixed Hamiltonian, the coefficients α and β can be obtained from an uniform transformation

$$h = \begin{pmatrix} \langle\Psi|\hat{H}|\Psi\rangle & \langle X|\hat{H}|\Psi\rangle \\ \langle\Psi|\hat{H}|X\rangle & \langle X|\hat{H}|X\rangle \end{pmatrix} \qquad (2.39)$$

The uniform transformation can be expressed in terms of a rotation matrix [61]

$$U = \begin{pmatrix} \cos\theta & \sin\theta \\ -\sin\theta & \cos\theta \end{pmatrix} \qquad (2.40)$$

with θ being

$$\tan\theta = \frac{1}{2}\frac{\mathrm{Re}\{h_{10}\} + \mathrm{Re}\{h_{01}\}}{\mathrm{Re}\{h_{00}\} + \mathrm{Re}\{h_{11}\}}. \qquad (2.41)$$

The final expression for improving the wave function reads then

$$|\Psi^{(n)}\rangle = U_{00}|\Psi^{(n-1)}\rangle + U_{01}|X^{(n)}\rangle. \qquad (2.42)$$

This scheme determines an angle θ which yields a wave function closest to the next *extreme* value of the corresponding one-particle energy. Only for a rather accurate initial guess of ϱ_{in} and $|\Psi^{(0)}\rangle$ it is ensured that this extreme value is a minimum of ε. However, if θ returns a maximum value, an addition of $\frac{\pi}{2}$ yields a minimum again.

So far the charge density is kept fixed. In order to obtain self-consistency an outer iteration can be applied which diagonalizes $\hat{H}$ at the updated charge density. The resulting wave functions Ψ yields the output density

$$\varrho_{\mathrm{out}}^{(n)}(\mathbf{R}) = \sum_{i\sigma\mathbf{k}} \omega_{\mathbf{k}} f_{i\sigma\mathbf{k}}^{\mathrm{occ}} |\Psi_{i\sigma\mathbf{k}}^{(n)}(\mathbf{R})|^2. \qquad (2.43)$$

Note, that the output density $\varrho_{\mathrm{out}}^{(n)}$ is *not* self-consistent anymore! This non-self-consistent density $\varrho_{\mathrm{out}}^{(n)}$ must not be used as input density $\varrho_{\mathrm{in}}^{(n+1)}$ in the subsequent diagonalization of $\hat{H}[\varrho_{\mathrm{in}}^{(n+1)}]$. Otherwise, the non-self-consistent density would introduce an increase of the total energy - in other words, the algorithm would diverge!

Instead, a "most self-consistent" or optimal density $\varrho_{\mathrm{opt}}^{(n)}$ can be computed from both $\varrho_{\mathrm{in}}^{(n)}$ and $\varrho_{\mathrm{out}}^{(n)}$. Therefore a functional $\varrho_{\mathrm{opt}}[\varrho_{\mathrm{out}}, \varrho_{\mathrm{in}}]$ has to be found which yields a maximum total energy gain in the next diagonalization step. The simplest approach is a linear mixer (also known as Pratt mixer [95])

$$\varrho_{\mathrm{opt}}^{(n)} = \alpha \varrho_{\mathrm{out}}^{(n)} + (1-\alpha)\varrho_{\mathrm{in}}^{(n)}. \qquad (2.44)$$

with α being a (constant) mixing factor between 0 and 1. This mixing scheme suffers usually from a bad convergence. Additionally, the mixing parameter α has to be optimized for each system.

---

[5] Because the charge density is fixed the contributions to the effective potential v_{eff} do not need to be recomputed. Only the non-local potential v_{nl} must be updated for each gradient calculation.

RMM-DIIS Beside the linear charge density mixer in this work the more advanced Pulay scheme [96] will be implemented. They key entity for charge density mixing schemes is the residual vector

$$R^{(n)}[\varrho] = \varrho_{\text{out}}^{(n)}[\varrho_{\text{in}}^{(n)}] - \varrho_{\text{in}}^{(n)}. \tag{2.45}$$

Convergence is reached when R is (nearly) zero. Thus, a minimization of $R[\varrho] \to 0$ can be used in order to estimate ϱ_{opt}.

Instead of using just $\varrho_{\text{in}}^{(n)}$ and $\varrho_{\text{out}}^{(n)}$ a set of previous densities could be used in order to predict $\varrho_{\text{opt}}^{(n)}$. Assuming a linear dependence between the previous densities the new "most self-consistent" charge density can be expressed as a linear combination

$$\varrho_{\text{opt}}^{(n)} = \sum_j \alpha^{(j)} \varrho_{\text{in}}^{(j)}. \tag{2.46}$$

The scheme to obtain the Pulay coefficients α will be introduced below. Here the optimal density ϱ_{opt} lies in the subspace spanned from the previous $\varrho_{\text{in}}^{(j)}$.

Hence, in each new iteration $\varrho_{\text{in}}^{(n+1)}$ has to introduce new variations of the density in order to extend that subspace. The procedure is known as DIIS[6] and was suggested by Pulay [96]. Since the DIIS mixing schemes assume linearity between $\varrho_{\text{in}}^{(n)}$ and $\varrho_{\text{out}}^{(n)}$ they have explicitly to be "decoupled" from $|\Psi\rangle$ by using a subspace diagonalization. The subspace diagonalization ensures $\varrho_{\text{opt}}^{(n)}$ to be the only solution and $\varrho_{\text{in}}^{(n)}$ and $\varrho_{\text{out}}^{(n)}$ are no longer linear dependent.

Knowing the self-consistent solution ϱ_{scf} the missing density $\varrho_{\text{in}}^{(n)} - \varrho_{\text{scf}}$ could be considered as a perturbation $\hat{v}_{\text{Coul}}[\varrho_{\text{in}}^{(n)} - \varrho_{\text{scf}}]$. Linearizing the residual vector R at the self-consistent density ϱ_{scf} gives

$$R[\varrho_{\text{in}}^{(n)}] \approx \mathbf{J}(\varrho_{\text{in}}^{(n)} - \varrho_{\text{scf}}) \tag{2.47}$$

with $\mathbf{J}$ being the dielectric function

$$\mathbf{J} = \mathbf{1} - \chi \mathbf{U} = \mathbf{1} - \chi \delta_{\mathbf{GG'}} \frac{4\pi e^2}{|\mathbf{G}|^2} \tag{2.48}$$

and χ being the susceptibility. It can be seen that the $\frac{1}{|\mathbf{G}|^2}$ term of the Hartree potential (see also Eq. (1.65)) introduces numerical errors[7] for small $|\mathbf{G}|$ values, which happens in case of large lattice vectors. These numerical errors cause instabilities in the gradient and eventually, in the long wavelength limit of the residual vector. This leads to artificial oscillations between subsequent charge densities $\varrho_{\text{in}}^{(n)}$. This effect is known as *charge sloshing*.

Similarly to the considerations on preconditioners in Sec. 2.1.3 it is again the condition of the dielectric matrix $\mathbf{J}$ which determines the convergence. The larger the ratio between largest and smallest eigenvalue the worse the convergence rate.

During the minimization the exact dielectric matrix $\mathbf{J}$ remains unknown. Its direct solution would be too expensive. However, an approximation of $\mathbf{J}$ (or $\mathbf{J}^{-1}$) can be used in order to describe the influence of the Hartree potential. Kerker [97] introduced an approximation to the response function from the Thomas-Fermi-screening. The metric reads

$$\mathbf{J}^{-1} \approx \delta_{\mathbf{GG'}} \frac{|\mathbf{G}|^2}{|\mathbf{G}|^2 + q_0}. \tag{2.49}$$

---

[6]DIIS = Direct Inversion of the Iterative Subspace
[7]In general, it is numerically unstable to divide by small numbers.

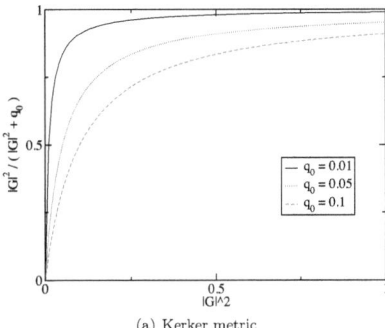

(a) Kerker metric

Figure 2.5: Kerker metric. Low **G** -frequency contributions are suppressed in the residual vector. Note, the **G**=0 component of R is zero anyway. The **G**=0 component carries the norm. In case of charge density *differences* that norm vanishes [71].

The frequency range that should be suppressed in the residual vector can be tuned with q_0. The shape of the Kerker metric is plotted in Fig. 2.5 for typical values of q_0.

Using[8] **J**-**1R** instead of **R** can counteract the charge sloshing [98]. In case of semiconducting/insulating systems this condition becomes a constant. For metallic systems there is no screening in the short wavelength limit (**J** $\approx$ **1**) whereas in the long wavelength limit the condition gets **J** $\approx 1/q^2 \sim L^2$ (with the metallic screening L). Hence, metallic systems are sensitive for charge sloshing problems and tend to converge significantly slower than semiconductors/insulators.

Combining Kerker's metric with Pulay's DIIS scheme results in the RMM-DIIS[9] charge density mixer. With the differences of the residual vectors

$$\Delta R^{(n)} = R^{(n)} - R^{(n-1)} \tag{2.50}$$

the Pulay matrix

$$A_{ij} = \langle \Delta R^{(i)} | \Delta R^{(j)} \rangle \tag{2.51}$$

can be constructed [28]. The Pulay coefficients $\alpha^{(n)}$ from Eq. (2.46) can be computed by inverting **A**

$$\alpha^{(n)} = -\mathbf{A}^{-1}\mathbf{B} \quad \text{with} \tag{2.52}$$

$$B_n = \langle \Delta R^{(n)} | R^{(m)} \rangle. \tag{2.53}$$

The optimal charge density for the next iteration $m+1$ is then

$$\varrho_{\text{opt}}^{(m+1)} = KR^{(m)} + \sum_{i}^{m-1} \alpha^{(i)}((\varrho_{\text{in}}^{(i)} - \varrho_{\text{in}}^{(i-1)}) + K\Delta R^{(i)}). \tag{2.54}$$

---
[8] **1** denotes the identity matrix.
[9] RMM-DIIS = Residual Metric Minimization - Direct Inversion of the Iterative Subspace

2.2 Structural properties

Besides describing electronic properties, the computation of structural and thermodynamic properties is important in CMD (such as the relaxed equilibrium atomic geometry, transition state search, and molecular dynamics). Since the S/PHI/nX framework should be able to compute such important material properties will briefly sketch important structural methods.

With the Hellmann-Feynman theorem (and the Pulay corrections) a way is found to determine forces efficiently within DFT. With an access to forces the relaxed atomic structure can be computed, transition states can be identified, and dynamic properties can be expressed. In the following paragraphs we present the basic concepts behind these schemes.

2.2.1 Quasi Newton

With the obtained forces the atomic structure of a system can be relaxed. One of the most efficient structural relaxation schemes is the quasi-Newton scheme which will be implemented in S/PHI/nX. Quasi means to avoid exploiting the actual Hessian (as used in all Newton-based schemes). The class of iterative quasi Newton algorithms use in each cycle an approximation of the inverse Hessian $\mathbf{H}^{-1}$ called $\tilde{\mathbf{B}}$. In modern applications the BFGS[10][99] quasi Newton scheme is one of the most successful optimization algorithms as it is extremely efficient for large-scale problem. Several variations of this algorithm have been developed to reach also $O(n)$ scaling.

In S/PHI/nX the BFGS quasi Newton scheme has been implemented. As all quasi-Newtons schemes it starts from an approximation of the Hessian. As initial guess the identity $\mathbf{1}$ is used as approximation

$$\mathbf{B} \approx \mathbf{H} = -\mathbf{1} \tag{2.55}$$
$$\tilde{\mathbf{B}} \approx \mathbf{H}^{-1}. \tag{2.56}$$

The gradient $\mathbf{g}$ can be identified with the forces $\mathbf{f}$ acting on each atom. The forces are obtained from either the Hellmann-Feynman forces from the DFT potential or analytically/numerically from empirical potentials. The gradient can be obtained from

$$\mathbf{g} = \mathbf{f} = \frac{\delta E}{\delta \tau}. \tag{2.57}$$

The atomic positions τ are improved along the gradient taken the curvature into account

$$\tau' = \tau - \tilde{\mathbf{B}}\mathbf{g}.$$

The gradient $\mathbf{g}'$ at the improved coordinates τ' reads

$$\mathbf{g}' = \mathbf{f}' = \frac{\delta E}{\delta \tau'}.$$

---
[10]BFGS = Broyden, Fletcher, Goldfard, Shanno

The BFGS algorithm updates the approximation of the inverse Hessian like

$$\tilde{\mathbf{B}}' = \left(1 - \frac{\mathbf{sy}^T}{\mathbf{y}^T\mathbf{s}}\right)\tilde{\mathbf{B}}\left(1 - \frac{\mathbf{ys}^T}{\mathbf{y}^T\mathbf{s}}\right) + \frac{\mathbf{ss}^T}{\mathbf{y}^T\mathbf{s}} \quad (2.58)$$

$$\mathbf{s} = \Delta\tau = \tau' - \tau \quad (2.59)$$

$$\mathbf{y} = \Delta\mathbf{g} = \mathbf{g}' - \mathbf{g}. \quad (2.60)$$

The approximation of the inverse Hessian is iteratively improved [100]

$$\tilde{\mathbf{B}}^{(n+1)} = \tilde{\mathbf{B}}^{(n)} - \frac{\tilde{\mathbf{B}}^{(n)}\mathbf{ss}^T\tilde{\mathbf{B}}^{(n)}}{\mathbf{s}^T\mathbf{Bs}} - \frac{\mathbf{yy}^T}{\mathbf{y}^T\mathbf{s}}. \quad (2.61)$$

2.2.2 Molecular dynamics

Properties which are dynamic in time, such as melting/solidification behavior, analysis of diffusion paths and diffusion barriers, and the study of chemical reactions, require the description of the atom trajectories. Therefore, Newton's equation of motion Eq. (1.108) needs to be solved. In order to obtain the trajectories an integrator over the time is introduced. For different statistical ensembles different integrator schemes are applied. The Verlet integrator [101, 102] is commonly used for micro-canonical ensembles, the Nosé-Hoover integrator [103] for canonical ensembles.

Verlet integrator One of the most commonly used integrators in the micro-canonical ensemble is the Verlet integrator. It is known to be stable and provides time-reversibility. The atomic positions are expanded into two third order Taylor series, one forward in time, one in reverse time

$$I: \quad \tau(t+\Delta t) = \tau(t) + \dot\tau(t)\Delta t + \frac{1}{2}\ddot\tau(t)\Delta t^2 + \frac{1}{6}\dddot\tau(t)\Delta t^3 + O(\Delta t^4) \quad (2.62)$$

$$II: \quad \tau(t-\Delta t) = \tau(t) + \dot\tau(t)\Delta t + \frac{1}{2}\ddot\tau(t)\Delta t^2 - \frac{1}{6}\dddot\tau(t)\Delta t^3 + O(\Delta t^4). \quad (2.63)$$

Adding (I) and (II) yields the actual Verlet integrator

$$\tau(t+\Delta t) = 2\tau(t) - \tau(t-\Delta t) + \Delta t^2 \ddot\tau + O(\Delta t^4). \quad (2.64)$$

The truncation error is of order $O(\Delta t^4)$. The third order term of the Taylor expansions cancels out in the Verlet integrator.

Subtracting (I) and (II) provides an expression for the velocities:

$$v(t) = \frac{\tau(t+\Delta t) - \tau(t-\Delta t)}{2\Delta t} + O(\Delta t^2). \quad (2.65)$$

In contrast to the atomic positions the velocities are only of order $O(\Delta t^2)$.

In principle, a more sophisticated algorithm such as higher order finite differences could be employed to reduce the integration error beyond $O(\Delta t^4)$ to allow a longer time step. A longer time step would reduce the computational time because the number of integration steps would be reduced. However, higher order terms do not significantly increase the maximum stable time step. On the other hand, higher order terms require both memory and additional computational effort compared to the plain Verlet integrator[11]. This is the reason why in the realm of the micro-canonical ensemble the Verlet algorithm is very popular.

Nosé-Hoover thermostat While in calculations in the micro-canonical ensemble the energy E, the volume V, and the particle number N is being kept constant (EVN ensemble) in the canonical ensemble the temperature T is kept fixed instead of the energy (TVN ensemble). The constant temperature constraint is usually accomplished within a Nosé-Hoover thermostat [103]. Here an extended Lagrangian is being introduced

$$L^{\text{Nose}} = \sum_{\tau}^{n_{\text{at}}} \frac{\mathbf{p}_\tau^2}{2M_\tau s^2} - E^{\text{BOS}}(\{\tau\}) + \frac{1}{2}Q\dot{s}^2 - \frac{L}{\beta}.$$

In this equation Q denotes an artificial mass of the additional degree of freedom. L is the number of physical degrees of freedom and the reciprocal temperature is β. The momentum $\mathbf{p}_\tau$ is given as

$$\mathbf{p}_\tau = \frac{\partial L^{\text{Nose}}}{\partial \tau} = M_\tau s^2 \dot{\tau}. \tag{2.66}$$

The Nosé Lagrangian leads to an extended Hamiltonian

$$\hat{H}^{\text{Nose}} = \sum_\tau \frac{\mathbf{p}_\tau^2}{2M_\tau} + E^{\text{BOS}}(\{\tau\}) + \frac{1}{2}Q + L\frac{\ln s}{\beta}.$$

2.3 Deriving thermodynamic properties

The S/PHI/nX framework will be applied in this work to compute thermodynamic properties (the phonon spectra $\omega(T)$, the linear expansion coefficients $\alpha(T)$, the heat capacities $C_V(T)$ and $C_p(T)$) of III-V semiconductors. In the following paragraphs a method to compute these data from first-principles is sketched.

2.3.1 Free energy surface

Thermodynamic properties of crystals can be derived from the free energy surface $F(T, V)$ spanned by the temperature T and the volume V. In case of non-magnetic crystals the free energy is decomposed into an electronic and a vibronic contribution [52]

$$F(T, V) = F^{\text{el}}(T, V) + F^{\text{vib}}(T, V). \tag{2.67}$$

The latter equation holds when the adiabatic approximation applies and higher order contributions (e.g., due to spin-orbit coupling) are negligible. The electronic contribution to the free energy F^{el} can be obtained in the finite temperature DFT [104] from the total energy E_{tot} of the crystal and the electronic entropy S^{el}

$$F^{\text{el}}(T, V) = E_{\text{tot}}(V) - TS^{\text{el}}, \tag{2.68}$$

---

[11] Memory and computational efficiency is not strictly relevant for MD implementations acting on DFT potentials since the computational effort of a MD algorithm is negligible compared to the computation of the Born-Oppenheimer surface. However, our framework should also support (semi-)empirical potentials which are computationally less demanding. In this case efficiency considerations of the MD implementation become important.

while the electronic entropy S^{el} can be determined from the occupancies f_i^{occ} as

$$S^{\text{el}} = 2k_B \sum_i (f_i^{\text{occ}} \ln f_i^{\text{occ}} + (1 - f_i^{\text{occ}}) \ln(1 - f_i^{\text{occ}})) \qquad (2.69)$$

with the Boltzmann constant k_B [52]. The prefactor 2 accounts for the spin-degeneracy of each quasi-particle state. Note that f^{occ} is temperature dependent via Eq. (1.43). In this work we focus on semiconducting systems for which we will show (see Sec. 4.1) that the temperature dependence of TS^{el} plays only a minor role.

The vibronic contribution of Eq. (2.67) reads [105]

$$F^{\text{vib}}(V,T) = \frac{1}{n_{\text{at}}} \sum_i^{3n_{\text{at}}} \left(\frac{1}{2}\hbar\omega_i + k_B T \ln\left(1 - e^{-\frac{\hbar\omega_i}{k_B T}}\right) \right). \qquad (2.70)$$

Here the phonon frequencies ω_i are eigenvalues of the dynamical matrix $\mathbf{D}$

$$\mathbf{D}(\mathbf{q})\mathbf{v}_i(\mathbf{q}) = \omega_i(\mathbf{q})\mathbf{v}_i(\mathbf{q}) \qquad (2.71)$$

and $\mathbf{v}_i$ are the corresponding eigenvectors at the wave-vector $\mathbf{q}$. The dynamical matrix can be expressed in reciprocal space [37]

$$D_{\mu\nu}(\mathbf{q}) = \frac{1}{M} \sum_{i_\alpha i_\beta}^{n_{\text{at}}} \mathbf{F}_{\mu\nu}^{\text{IFC},i_\alpha i_\beta} e^{i\mathbf{q}\cdot\mathbf{R}}. \qquad (2.72)$$

$\mathbf{F}^{\text{IFC}}$ denotes the matrix of the interatomic force constants [106]

$$\mathbf{F}_{\mu\nu}^{\text{IFC},i_\alpha i_\beta} = \frac{\partial^2 F^{\text{el}}}{\partial \Delta\tau_{i_\alpha\mu} \partial \Delta\tau_{i_\beta\nu}} \qquad (2.73)$$

which is generated by an atomic displacement $\Delta\tau_{i_\alpha\mu}$ of atom i_α along the direction μ. This displacement induces forces on the atoms i_β along the directions ν.

There are different approaches to compute interatomic force constants. Originally, this has been done by inverting the dielectric matrix [107] or, computationally much less demanding, using a perturbation method using the Sternheimer equation [52] in the Density Functional Perturbation Theory (DFPT) [106, 108]. Alternatively, the interatomic force constants can be obtained using the direct method. The phonon frequencies are obtained from total energy differences of the unperturbed and the perturbed structure (frozen phonon method [1]) or, as used in this work, from forces acting on atoms in the distorted geometry using HF theorem (see Sec. 1.7.1).

Perturbative methods are better suited for strongly localized phonon anomalies and the direct method for shallow ones [109]. The direct method can be applied without the implementation of a perturbative Hamiltonian and can be used with the current version of S/PHI/nX.

For systems for which the temperature dependence of TS^{el} plays only a minor role (see above), the electronic contribution F^{el} can be identified with the total energy at the equilibrium volume V_{eq}

$$F^{\text{el}}(V,T) \cong E_{\text{tot}}(V_{\text{eq}}). \qquad (2.74)$$

Within DFT the dynamical matrix can be determined by evaluating the interatomic force constants using the Hellmann-Feynman theorem. With Eq. (2.74) $\mathbf{F}^{\text{IFC}}$ becomes [106]

$$\mathbf{F}^{\text{IFC}}_{i_\alpha\mu,i_\beta\nu} = \frac{\partial^2 E_{\text{tot}}}{\partial \Delta \tau_{i_\alpha\mu} \partial \Delta \tau_{i_\beta\nu}}. \tag{2.75}$$

Following Ref. [37] the volume dependent total energy $E_{\text{tot}}(V)$ can be obtained by constructing a fit to the Murnaghan equation

$$E_{\text{tot}}(V) = E_{\text{tot}}(V_{\text{eq}}, T=0) + \frac{BV}{B'^2 - B'}\left(B'(1 - \frac{V_{eq}}{V}) + \left(\frac{V_{eq}}{V}\right)^{B'} - 1\right) \tag{2.76}$$

with the bulk modulus B and its derivative B'. The free energy surface can then be computed as

$$F(T,V) = E_{\text{tot}}(V) + \frac{1}{n_{\text{at}}} \sum_i^{3n_{\text{at}}} \left(\frac{1}{2}\hbar\omega_i + k_{\text{B}}T \ln\left(1 - e^{-\frac{\hbar\omega_i}{k_{\text{B}}T}}\right)\right). \tag{2.77}$$

The thermodynamic properties $\omega(\mathbf{q})$, $\alpha(T)$, and $C_{p,V}(T)$ will be computed in this work from first-principles by combining Eqs. (2.71), (2.72), (2.75), (2.76), and (2.77).

Thermodynamic properties The free energy surface defined in Eq. (2.77) provides access to various thermodynamic properties [105], such as the thermal expansion ϵ and its coefficients α, constant volume heat capacity C_V and constant pressure heat capacity C_p or the mode-Grüneisen parameters γ

$$P = -\left(\frac{\partial F(T,V)}{\partial V}\right)_{V_{\text{eq}}} \tag{2.78}$$

$$\epsilon(T) = \frac{a(T) - a(T_{\text{ref}})}{a(T_{\text{ref}})} \tag{2.79}$$

$$\alpha(T) = \frac{1}{a(T)}\frac{\partial a(T)}{\partial T} \tag{2.80}$$

$$C_V = T\left(\frac{\partial^2 F(T,V)}{\partial T^2}\right)_{VV} \tag{2.81}$$

$$C_p = T\left(\frac{\partial^2 F(T,V)}{\partial T^2}\right)_{Vp} \tag{2.82}$$

$$\gamma = -\frac{V}{3n_{\text{at}}}\sum_i^{3n_{\text{at}}} \frac{1}{\omega_i}\frac{d\omega_i}{dV}. \tag{2.83}$$

The thermal expansion $\epsilon(T)$ is obtained from the lattice constant $a(T)$ and the lattice constant at a reference temperature T_{ref}.

2.3.2 Born-effective charges

In polar crystals, such as the zincblende III-V semiconductor systems we investigate in this study, long range electric fields emerge due to long-range longitudinal phonons. This effect is responsible for the disappearance of the degeneracy between longitudinal and transversal optical phonon (LO and TO) at the center of the Brillouin zone, also known as the LO-TO splitting phenomenon. Following Ref. [110], the origin of the LO-TO-splitting can be easily understood when considering an optical phonon of a polar zincblende crystal

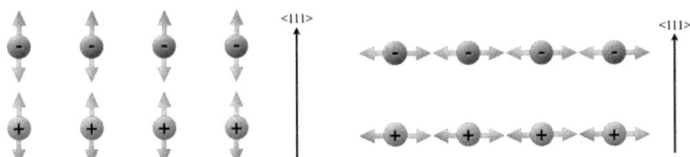

Figure 2.6: Sketch of (a) LO and (b) TO phonon modes.

along the $\langle 111 \rangle$ direction. In this orientation the positive and negative ions lie in separate parallel planes. In the LO phonon mode the ions are vibrating perpendicular to these planes (see Fig. 2.6). This is equivalent to a capacitor with oppositely charged plates sliding apart from each other inducing an extra force onto the plates. Analogously, due to the Coulomb interaction of the moving atomic planes perpendicular to the spatial diagonal an extra force is introduced. This additional force causes a frequency shift of the corresponding phonon. In the TO phonon mode the atoms vibrate within these planes. That is similar to a capacitor with infinite parallel plates which are sliding along the plates while keeping the distance between the plates constant. In contrast to the LO mode, here no additional force is induced and the phonon frequency is not being affected. From the above picture it also becomes clear that the frequency shift depends on the direction from where one approaches Γ. Depending on the approaching direction the distribution of the oppositely charged ions differs and thus, the induced frequency shifts.

The couplings between the optical phonons and the electric fields are called Born effective charges Z^*

$$Z^*_{\kappa,\beta\alpha} = \Omega \frac{\partial P_\beta}{\partial \tau_{\kappa\alpha}(\mathbf{q}=\mathbf{0})} \tag{2.84}$$

being the change of the macroscopic polarization P along direction β caused by an atomic displacement along direction α with absent external fields. With the electric enthalpy

$$E_{\text{electr}} = E - \Omega \sum_\alpha P_\alpha \varepsilon_\alpha \quad \text{and} \quad P_\alpha = -\frac{1}{\Omega} \frac{\partial E_{\text{electr}}}{\partial \varepsilon_\alpha} \tag{2.85}$$

the Born effective charge can be expressed in terms of the total energy or the forces as

$$Z^*_{\kappa,\alpha\beta} = -\frac{\partial^2 E_{\text{electr}}}{\partial \varepsilon_\beta \partial \tau_{\kappa\alpha}} = \left.\frac{\partial F_{\kappa\alpha}}{\partial \varepsilon_\beta}\right|_{\tau_{\kappa\alpha}=0}. \tag{2.86}$$

This equation is equivalent to the above discussion. The induced charge is due to the change of the force along α when considering a homogeneous electric field along β.

The coupling between phonons and electric fields affect the computed optical phonons near Γ [40]. The LO-TO splitting affects only a very small region of the $\mathbf{q}$ space. In this work we neglect the LO-TO splitting.

Conclusions

We presented the required theoretical and methodological concepts which are necessary to implement a flexible framework to describe problems in CMD. We introduced various theories to describe ground state properties such as density functional theory, (density functional based) tight binding, or the application of empirical potentials. A brief overview on methods to obtain electronic, structural, and thermodynamic properties based on these theories has been presented. With this knowledge a framework can be derived which is sufficiently flexible to allow an implementation of state-of-the-art algorithms to compute a wide range of material properties computationally efficiently and accurately.

Chapter 3

S/PHI/nX

Method development in the field of CMD is typically very time consuming. The implementation of new basis-sets, minimizers, or structural algorithms can easily create workloads of months or years. A major part of the time is spent in an iterative process of code implementation, testing/debugging, and a step to optimize the run-time performance of the code. The developer must typically have deep programming skills and a thorough understanding in numerics, computer science, and physics before an implementation can begin.

In the first step of the development process (code implementation) the physical/algebraic algorithms have to be transcribed to source code. The developer is often confronted with procedural programming leading to function call with 20 arguments and more. For example, a typical source code reads[1]:

```
call opernla_ylm(choice,dgxdt,ffnlin,gx,ia3,idir,istwf_k,&
   itype,kgin,kpin,lmnmax,matbtl,mincat,mlmn,ndgxdt,nincat,nloalg,&
   npwin,ntype,ph3din,vectin)
call opernc_ypm(atindx1,choice,dgxdt,dgxdtfac,dimenl1,dimenl2,&
   enl,gx,gxfac,iatm,indlmn,itype,lmnmax,mincat,mlmn,natom,ndgxdt,&
   ndgxdtfac,nincat,ntype,paw,signs,wt)
```

In order to introduce changes in such a code a detailed knowledge of available functions, and the exact meaning of all arguments is required. Required tasks like memory management and the numerical details do not allow the developer to completely focus on the actual algorithm. This slows down the implementation process significantly.

The testing and debugging sequence occupies typically significantly more time. Many program packages are not strictly modularized. Modifications in one part of the code can often introduce unwanted side effects in other routines. Locating such problems is often challenging. Furthermore, the explicit memory management can lead to unpredictable run-time behavior. Therefore, the time needed for testing and debugging is often the major part in code development.

Once the algorithm has been successfully transcribed and implemented the code needs to be optimized in order to run efficiently on modern computers. This step requires a thorough understanding in numerics, computer science, and computer hardware. New trends in HPC such as massive multi-core architectures,

---
[1] Taken from `Abinit/Src_2nonlocal/nonlop_ylm.f`.

multi-threaded programming, CUDA[2]/GPU[3] computing, grids/clouds have to be exploited when optimizing algorithms to take advantage from the new hardware capabilities. The setup of data structures, communications between computer components, and the efficient usage of external libraries need to be considered. It will be shown later that often only a fraction of the possible algorithmic optimization is reached in many HPC program packages.

This work introduces a new concept of how CMD method development can be simplified. We aim at a method which requires only rudimental programming skills from the physicists which allows to focus on the implementation of the actual algorithm. Therefore, we introduced a physics meta-language which allows the development of even complex quantum mechanical algorithms in the Dirac notation directly in the source code.

In this chapter the techniques are presented which we developed in the scope of this work. We use our new meta-language to create the efficient library and full-featured program package S/PHI/nX. It will be demonstrated that in our approach the implementation of a DFT Hamiltonian using a plane-wave basis-set can be accomplished in a very short and transparent source code (approx. 550 source lines). For example, a quantum mechanical expression such as

$$\Psi(\mathbf{R}) = \sum_{\mathbf{G}} \langle \mathbf{R}|\mathbf{G}\rangle \langle \mathbf{G}|\Psi\rangle$$

can be programmed in an almost text book like notation:

```
psiR = SUM (G, (R|G) * (G|psiG));
```

The entire approach has been derived under the constraint of achieving a very high run-time performance on modern computer platforms with a source code which is as close as possible to mathematics/physics textbooks. Therefore, we discuss our new techniques in an order dictated by performance considerations:

- We begin with an analysis of the machine code generated by the common programming languages in HPC (FORTRAN77, Fortran90/95, C, and C++) in order to identify the best suiting programming language for this project.

- Quantum mechanical expressions can be efficiently expressed in an algebraic formalism [52] which requires an efficient numeric/algebra library in our approach. We derive new techniques addressing automatic memory management and show how the run-time performance of *functional* programming[4] can be significantly increased. This allows us an implementation of a high performance algebra library with an interface reminiscent to algebraic textbooks.

- We discuss techniques which we developed to create a quantum-mechanical meta-language and describe how to get from a conventional modular programming via object-orientation to a functional Dirac-notation meta-language.

- We extend the set of techniques such that complex equations of motion can be comfortably implemented.

---
[2]Compute Unified Device Architecture. CUDA is a parallel computing architecture developed by NVIDIA
[3]Graphic processing unit.
[4]Functional programming, i.e., defining algorithms in terms of mathematical functions, is necessary for a notation resembling text books.

- The chapter is completed by run-time performance benchmarks of S/PHI/nX in comparison to VASP. Here we demonstrate that with our approach textbook-like source code can generate highly efficient CMD programs.

3.1 Basis-set independent implementations

Modern physics offers two convenient notations for quantum mechanical equations: the matrix-oriented Heisenberg notation (suggested by Heisenberg, Born, and Jordan [111]) and Dirac's vector-oriented bra-ket notation [35]. In the following we compare both styles with respect to the feasibility for developing a meta-language for quantum mechanics.

3.1.1 Matrix notation

When performing numerical calculations the wave functions can be expanded into basis functions $\{b_\alpha\}$

$$\Psi_i = \sum_\alpha C_{\alpha i} b_\alpha. \tag{3.1}$$

Here, the $C_{\alpha i}$ denote the expansion coefficients of the wave function Ψ_i. Thus, for numerical calculations the wave function can be treated as the matrix $C_{\alpha\beta}$

$$\mathbf{C} = \begin{pmatrix} C_{11} & C_{12} & \cdots & C_{1m} \\ C_{21} & C_{22} & \cdots & C_{2m} \\ \vdots & \vdots & \ddots & \vdots \\ C_{n1} & C_{n2} & \cdots & C_{nm} \end{pmatrix} \tag{3.2}$$

with each state/band i being a single column of that matrix. The dimension of the coefficient matrix is determined by n basis functions and m states.

Quantum mechanics makes heavy usage of operators. An arbitrary hermitian operator $\hat{A}$ which acts on Ψ can also be expressed as a matrix

$$\begin{pmatrix} C'_{11} & C'_{12} & \cdots & C'_{1m} \\ C'_{21} & C'_{22} & \cdots & C'_{2m} \\ \vdots & \vdots & \ddots & \vdots \\ C'_{n1} & C'_{n2} & \cdots & C'_{nm} \end{pmatrix} = \begin{pmatrix} A_{11} & A_{12} & \cdots & A_{1n} \\ A_{21} & A_{22} & \cdots & A_{2n} \\ \vdots & \vdots & \ddots & \vdots \\ A_{n1} & A_{n2} & \cdots & A_{nn} \end{pmatrix} \begin{pmatrix} C_{11} & C_{12} & \cdots & C_{1m} \\ C_{21} & C_{22} & \cdots & C_{2m} \\ \vdots & \vdots & \ddots & \vdots \\ C_{n1} & C_{n2} & \cdots & C_{nm} \end{pmatrix}. \tag{3.3}$$

Besides the wave functions basis-set operators like the overlap $\mathbf{O}$ and the Laplacian $\mathbf{L}^5$ can also be defined as matrices

---
[5]We follow here the nomenclature presented in Ref. [67].

$$\mathbf{O}_{\alpha\beta} = \int d\mathbf{r} b_\alpha^*(\mathbf{r}) b_\beta(\mathbf{r}) \tag{3.4}$$

$$\mathbf{L}_{\alpha\beta} = \int d\mathbf{r} b_\alpha^*(\mathbf{r}) \nabla^2 b_\beta(\mathbf{r}), \tag{3.5}$$

where $\mathbf{O}_{\alpha\beta} = \delta_{\alpha\beta}$ for orthonormal basis-sets.

Using this set of operators and functions any Hamiltonian can be built up independently of the basis-set. For the sake of simplicity we demonstrate the kinetic operator in a matrix notation which reads

$$\mathbf{T} = -\frac{1}{2} \sum_i \int d\mathbf{r} \Psi_i^*(\mathbf{r}) \nabla^2 \Psi_i(\mathbf{r}) \tag{3.6}$$

$$= -\frac{1}{2} \sum_{i\alpha\beta} C_{\alpha i}^* L_{\alpha\beta} C_{\beta i} \tag{3.7}$$

$$= -\frac{1}{2} \mathrm{tr}(\mathbf{C}^\dagger \mathbf{L} \mathbf{C}). \tag{3.8}$$

This formulation can easily be applied to all other contributions of the Hamiltonian as well. It has been successfully applied in the DFT++ project of the Arias Research Group Initiative [67]. The computational advantage of this approach is, that blocked algorithms can be used and highly optimized matrix-matrix numeric libraries (BLAS[6], LAPACK[7]) can be applied which improves the executional performance drastically. On the other hand, depending on the basis-set the wave function and operator matrices might tend to be very large, perhaps even too large to keep them in the computer's memory. In the Arias approach a fallback to the memory friendly vector-vector implementation is not straightforward. Hence, the choice of a vector or matrix representation is a compromise between memory consumption and performance. Therefore, one goal in this work is to find a solution that supports vector and matrix representation equally, depending on the algorithmic and computational needs.

3.1.2 Bra-Ket notation

Besides the matrix-oriented Heisenberg notation nowadays the Dirac notation has become a standard language in quantum mechanics because it allows "hiding" the basis-set details and focusing on the physical content instead. Generally, a Dirac vector is symbolized as $|\Psi\rangle$, and is an element of an abstract Hilbert space which cannot be used for performing actual calculations. In order to define a wave function numerically a basis-set $\langle \mathbf{B}|$ is necessary on which the wave function can be projected (sampled) on, e.g.,

$$\Psi(\mathbf{B}) = \langle \mathbf{B}|\Psi\rangle. \tag{3.9}$$

How that expansion looks in detail is not specified in the *bra-ket* notation. The Dirac notation also allows projections between different basis-sets, e.g,

---

[6]BLAS = Basic Linear Algebra System. BLAS is a set of highly optimized algebra functions organized in BLAS-1 (vector-vector operations), BLAS-2 (vector-matrix operations), and BLAS-3 (matrix-matrix operations). BLAS libraries are typically provided by the HPC vendors and are specially optimized for their platforms, e.g., AMD-ACML, Intel-IMKL, IBM-Essl, HP-Veclib.

[7]LAPACK = Linear Algebra Package. LAPACK provides high level algebra algorithms such as matrix inversions or singular value decompositions based on BLAS.

$$\Psi(\mathbf{X}) = \sum_{\mathbf{B}} \langle \mathbf{X}|\mathbf{B}\rangle\langle \mathbf{B}|\Psi\rangle = \langle \mathbf{X}|\Psi\rangle. \tag{3.10}$$

An arbitrary hermitian operator $\hat{A}$ which acts on $|\Psi\rangle$ yields a vector $|\Psi'\rangle$

$$\hat{A}|\Psi\rangle = |\Psi'\rangle. \tag{3.11}$$

Multiplying with the adjoined *bra* vector gives the expectation values (matrix elements)

$$\langle \Psi_i|\hat{A}|\Psi_j\rangle = \langle \Psi_i|\Psi'_j\rangle. \tag{3.12}$$

In order to define a basis-set independent Dirac-like implementation, basis-set dependent operators like the overlap $\hat{\mathbf{O}}$ or a Laplacian $\hat{\mathbf{L}}$ have to be defined. Then, based on those basis-set dependent operators a new Hamiltonian can be built up which depends only on these operators.

A special operator is the density matrix

$$\hat{\varrho} = \sum_i |\Psi_i\rangle\langle \Psi_i| \tag{3.13}$$

which can be used to express expectation values according to

$$\langle \Psi_i|\hat{A}|\Psi_i\rangle = \mathrm{tr}(\hat{\varrho}\hat{\mathbf{A}}). \tag{3.14}$$

Of course, both formulation styles, the matrix and the Dirac notation, can easily be connected with each other. So the Dirac *ket* vector $|\Psi\rangle$ can be expressed as a column vector of its expansion coefficients

$$\langle \mathbf{B}|\Psi_i\rangle = \begin{pmatrix} C_{1i} \\ C_{2i} \\ \vdots \\ C_{ni} \end{pmatrix} \tag{3.15}$$

and its adjoined as a *bra* row vector, which is

$$\langle \Psi_i|\mathbf{B}\rangle = (C_{i1}^* C_{i2}^* \cdots C_{in}^*). \tag{3.16}$$

The scalar product of a *bra* and a *ket*[8] is then simply[9]

$$\langle \Psi'_i|\mathbf{B}\rangle\langle \mathbf{B}|\Psi_j\rangle \Rightarrow \sum_k C_{ik}^* C_{kj} = (C_{i1}^* C_{i2}^* \cdots C_{in}^*) \begin{pmatrix} C_{1j} \\ C_{2j} \\ \vdots \\ C_{nj} \end{pmatrix}. \tag{3.17}$$

---

[8]This is why the Dirac notation often is also referred to as *bra-ket* notation.
[9]This works only if both vectors are given in a complete orthonormal basis. Otherwise a metric has also to be introduced.

Such scalar products are massively used in quantum mechanics. Numerically they can be comfortably evaluated using BLAS function calls.

Most of the algorithms in literature are presented in the *bra-ket* formalism. If a programming language would support this notation implementation of quantum mechanical program packages would be dramatically simplified. In the following sections we derive the necessary techniques to extend existing programming languages such that a support of the flexible Dirac formalism becomes possible.

3.1.3 Programming languages

The support of the Dirac notation directly in a programming language is not trivial. In the Dirac formalism similar terms require the execution of very different operations depending on the physical context. For example, the meaning of the term $\langle \mathbf{A}|\mathbf{B}\rangle$ depends on what $\mathbf{A}$ and $\mathbf{B}$ is:

$$\langle \mathbf{X}|\Psi\rangle \Rightarrow \begin{cases} \text{identity} & \text{if } \Psi = \Psi(\mathbf{X}) \\ \text{projector from } \mathbf{B} \text{ to } \mathbf{X} & \text{if } \Psi = \Psi(\mathbf{B}) \end{cases} \qquad (3.18)$$

$$\langle \Psi|\Psi\rangle \Rightarrow \text{scalar product (with metric)} \qquad (3.19)$$

$$\langle \mathbf{X}|\mathbf{B}\rangle \Rightarrow \text{function call to projector between } \mathbf{B} \text{ and } \mathbf{X} \qquad (3.20)$$

$$\langle \hat{\mathbf{A}}|\Psi\rangle \Rightarrow \text{evaluation of an operator A such as } \hat{O}, \hat{L}, \hat{H} \qquad (3.21)$$

In order to assemble machine code, the compiler has to recognize the physical meaning. For a physicist/chemist this discrimination of the above cases is trivial while "teaching" the physical context to a compiler is challenging and requires the combination of various disciplines of computer science and quantum physics. A programming language has to be identified that provides the flexibility to allow an extension for "understanding" the basics of a quantum-mechanical language. Therefore, in this section we compare the common programming languages used in HPC with respect to the efficiency of the generated machine code and the provided flexibility in the source code.

Most of the scientific software so far has been written in FORTRAN77 or Fortran90/95. These languages generate fast machine code but do not provide the huge freedom to the developer as C or C++ do. Unfortunately, C and in particular C++ is slow in the field of numerics [112]. The slower executional speed of C++ is also referred to as *abstraction penalty*. The key in overcoming the abstraction penalties is the understanding of how both languages define pointers[10]. In the following a brief analysis of pointer arithmetics and memory hierarchies will be presented.

Performance of vector operations

FORTRAN represents vectors in form of arrays. When accessing the i-th vector element FORTRAN internally interprets it as a pointer to an address with the offset $n_{\text{bytes}} * i$ with n_{bytes} being the number of bytes used for storing one element, e.g., 8 bytes for a REAL*8. Here it is important that FORTRAN's pointer can dereference[11] only addresses that lay *inside* the range belonging to the vector. The FORTRAN compiler makes sure that the pointer cannot refer to any other address in the memory. Also the usage of

---
[10]Pointer = address in memory. A pointer to a variable "points to" the address of that variable.
[11]dereferencing = resolving the value at the memory address specified by the pointer.

Fortran90/95's pointer type does not allow referring to an address outside the assigned vector[12]. This *restricted* definition of a pointer has a big advantage for high-performance numerics: the compiler makes sure that two pointers cannot overlap. Such non-overlapping pointers are called *restricted* pointers. Assuming the vector-vector operation $\mathbf{y} = \mathbf{y} + a\mathbf{x}$, a FORTRAN source code would read

```
do i=1,n
    y[i] = y[i] + a*x[i]
end do ! i
```

Because the (internal) pointers of $y[i] = y_{\text{start}} + n_{\text{bytes}} i$ and $x[i] = x_{\text{start}} + n_{\text{bytes}} i$ cannot overlap by definition, the FORTRAN compiler will unroll the loop over i and split up the expression. During the optimization sequence the compiler generates an intermediate code like

```
do i=1, n, 4
    y[i]   = y[i]   + a*x[i]     ! performed in 1st CPU pipeline
    y[i+1] = y[i+1] + a*x[i+1]   ! performed in 2nd CPU pipeline
    y[i+2] = y[i+2] + a*x[i+2]   ! performed in 3rd CPU pipeline
    y[i+3] = y[i+3] + a*x[i+3]   ! performed in 4th CPU pipeline
end do ! i
```

The 4 assignments are completely independent from each other (because the pointers are restricted and cannot overlap). Thus, they can also be evaluated independently in the 4 parallel software-pipelines of the CPU[13]. As a result the executional speed increases roughly by a factor of 4!

In C/C++ the situation is more complicated because here the concept of pointers is by far more flexible. This flexibility allows the realization of object-oriented techniques. In C/C++ a pointer can always overlap with another pointer[14]. In the previous example the C/C++ compiler cannot assume that the 4 assignments are really independent ($\mathbf{x}$ can also point to some area of $\mathbf{y}$). In this case unrolling of the i loop would not lead to a parallel execution in the CPU's pipelines. However, with this knowledge in mind it is simple to overcome this problem manually by loading the values into local variables. The compiler always assumes variables to be independent from one another. The corresponding C code (which yields the same speed as the optimized FORTRAN code) reads

```
for (i=0; i < n; i+=4) {
    x0 = x[i+0]; x1 = x[i+1]; x2 = x[i+2]; x3 = x[i+3];
    y0 = y[i+0]; y1 = y[i+1]; y2 = y[i+2]; y3 = y[i+3];
    y[i]   = y0 + a*x0;   ! performed in 1st CPU pipeline
    y[i+1] = y1 + a*x1;   ! performed in 2nd CPU pipeline
    y[i+2] = y2 + a*x2;   ! performed in 3rd CPU pipeline
    y[i+3] = y3 + a*x3;   ! performed in 4th CPU pipeline
}
```

Of course, this manual unrolling can be implemented in low-level libraries and can so be hidden from the developer.

---

[12] The only exception is the NULL address to indicate an not associated pointer.

[13] The number of available pipelines depends on the architecture. We discuss here the situation for an Intel Xeon platform with 4 pipelines.

[14] Bjarne Stroustrup, who developed the C++ language, suggested a new pointer attribute **restrict** to support the usage of restricted pointers. However, the introduction of restricted pointers was not approved by the ANSI C++ committee. Hence, it is not supported by all C++ compilers and we, therefore, avoid it.

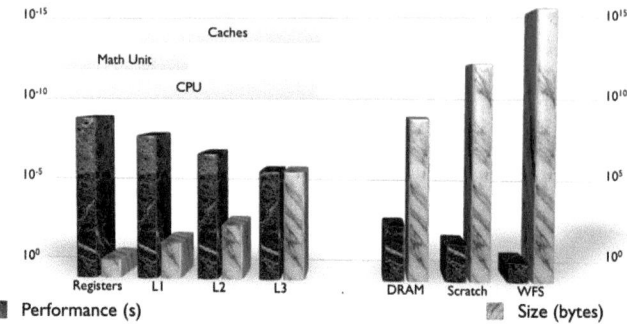

Figure 3.1: Memory hierarchy in modern computer architectures. Since most numerical operations have to reuse certain elements of the memory (such as matrix operations) a hierarchy of memories is nowadays implemented in all modern computer platforms. Fast memory is expensive and very limited in space. The fastest memory are CPU registers and Level 1 Caches located close to the math unit (access time: ns, size: 100 bytes). To buffer data from the RAM a hierarchy of 3 level caches is provided inside the CPU. The RAM is connected to the CPU via a (slow) data bus (100ns, GB). Scratch space refers to the local hard disk while WFS are global file systems mounted via a slow network. The massively accessed elements can be first loaded into the smaller but faster memory. This way the operations can be performed much faster. Depending on the number of available memory levels this approach can be recursively repeated.

Matrix operations

The other extremely important performance issue is the usage of the CPU's level caches[15] (see Fig. 3.1). Matrix operations can greatly benefit from such architectures. This can be easily demonstrated by means of a simple matrix multiplication $C_{ij} = \sum_k A_{ik} B_{kj}$. A pseudo code of a naive implementation would read

```
do i=1, n
  do j=1, n
    do k=1, n
      C[i,j] = C[i,j] + A[i,k] * B[k,j]
    end do ! k
  end do ! j
end do ! i
```

In this implementation the matrix multiplication involves $2n^3$ arithmetic operations if "+" and "*" are counted separately. The memory demand for the three matrices A, B, and C are in total as large as $3n^2$. If A, B, and C do not fit into the CPU's level caches the matrix elements of A and B have to be loaded repeatedly from the RAM through the slow data bus. If A and B are subdivided into smaller $b \times b$ matrices such that the fast level caches can accommodate the blocks the performance speed can be dramatically increased. An improved algorithm would read

```
do i=1, n
```

---

[15] The level caches are memory banks integrated directly inside or placed very close to the CPU. Due to the short distances to the math units accessing of the level cache memories is extremely fast compared to the slow access to the RAMs via the data bus. Modern CPUs are equipped with a hierarchical arrangement of caches, most common are 3 levels of different sizes, ranging from a few kB (level 1 cache, fastest access) to 2 MB (level 3 cache).

```
do j=1,n
   do k=1, n
      ! --- loop over blocks
      do iBlk=i, min(i+b-1, n)
         do jBlk=j, min (j+b-1, n)
            do kBlk=k, min(k+b-1,n)
               C[iBlk,jBlk] = C[iBlk,jBlk] + A[iBlk,kBlk]*B[kBlk,jBlk]
            end do ! kBlk
         end do ! jBlk
      end do ! iBlk

   end do ! k
end do ! j
end do ! i
```

This implementation results in exactly the same arithmetic operations as in the first example. Only the sequence differs between the unblocked and blocked version. Consider a single iteration at a fixed tuplet of (i,j,k). The $3b^3$ arithmetic operations that are performed in the inner loops operate on data blocks with the size $3b^2$. If b has been chosen small enough that the data blocks fit into the cache the data transfer between the slower RAM can be avoided. Depending on the computer platform, the speed up is up to several orders of magnitude! The standardized numeric libraries BLAS and LAPACK are highly focusing on such level cache techniques.

3.1.4 BLAS/LAPACK interfaces

Procedural vs. functional interfaces

An integral part of this thesis is the development of an intuitive programming interface that automatically translates algebraic expressions into highly optimized function calls exploiting the above described issues. BLAS and LAPACK provide highly efficient numeric subroutines, but using their interfaces is cumbersome, e.g., a multiplication of a double precision general matrix A and matrix B requires a subroutine call with 13 arguments:

```
call DGEMM (tranA, tranB, m, n, k, alpha, A, lda, B, ldb, beta, c, ldc)
```

We refer to the BLAS documentation for the meaning of the arguments. When optimizing a program package every algebra expression has to be replaced with such subroutine calls. In turn, the source code becomes very difficult to read and to maintain. Furthermore, the large amount of subroutine arguments in BLAS/LAPACK calls can slow down the development process. With various indices the developer might introduce inconsistent arguments which are often not detected by the subroutine. The runtime behavior can become unpredictable. The identification of the problem (often referred to as *debugging*) can be very challenging. Due to the cumbersome subroutine interfaces of BLAS/LAPACK in many program packages only key algorithms are optimized accordingly and the executional performance suffers.

In order to simplify the handling in many simulation packages wrapper procedures to the most important BLAS/LAPACK calls have been developed. Here, the additional information like dimensions are often hidden in modules/classes to reduce the number of arguments drastically, for example:

```
call MatrixMult (result, A, B)
```

Implementing algorithms based on such wrappers is much simpler than performing direct BLAS/LAPACK calls, but such handling it still not intuitive. Algebraic equations could be better expressed using a *functional* programming interface instead of calling subroutines. Therefore, some high-level programming languages (like Fortran90/95 or C++) support the definition of own data types in combination with overloaded operators. Using this technique unique data types for vectors or matrices can be defined which provide the proper BLAS/LAPACK wrapper procedures. Operators, like "*" can be overloaded to execute the actual BLAS/LAPACK DGEMM call. In program packages which provide such an algebra interface the above example can now easily be transcribed to

```
MyMatrix :: result, A, B
A = ...
B = ...
result = A * B
```

The multiplication operator is simply overloaded such as

```
MyMatrix operator* (Matrix A, Matrix B)
{
    Matrix result
    call MatrixMult (result, A, B)
    return result
}
```

The matrix elements can be of different data types, such as integer, single/double real, or single/double complex. That means that the operator function has to be provided for all combinations of possible data types. Furthermore, the actual BLAS/LAPACK interface depends on the matrix type (general, symmetric, hermitian, trigonal packed, etc.). Therefore, the multiplication function needs to support all combinations of data *and* matrix types. These combinations have to be considered for every algebraic operation. Although such an approach would offer an efficient and intuitive programming interface for algebraic expressions, the manual development of such a library would not be feasible.

In C++ functions can be defined as *templates*. In this programming technique functions can be developed once for all possible data types. The actual type will be kept as template argument, e.g., <T>. During the compilation, template arguments are replaced with proper data types. Hence, the compiler can create functions with all possible combinations of types[16]. A simplified C++ source code could read as follows:

```
template<class T>
MyMatrix<T> operator* (MyMatrix<T> a, MyMatrix<T> b)
{
    MyMatrix<T> result;
    MatrixMult (&result, a, b);
    return result;
}
```

---

[16]Since the compiler generates source code from generic types, such approach is often called *generic programming*.

This operator can be invoked for various data types:

```
MyMatrix<double> a = ..., b = ..., c;
MyMatrix<complex16> d = ...; e = ...; f;
c = a * b;
f = d * e;
```

Information about dimensions and matrix/vector element types are provided only during the declaration. The source code addressing the actual algorithm is kept free of memory management and BLAS/LAPACK function call mappings. Such an intuitive handling of algebraic expressions can speed up the development and debugging process significantly. The task of memory management and numeric function mapping is accomplished fully automatically *at compile time*. In contrast to a manual optimization, the compiler replaces *systematically* all algebraic expressions with highly efficient subroutines. A manually written algebra program can be as fast as such a generic library only if all numeric expressions have been replaced with proper calls thoroughly.

A functional programming ansatz has the huge advantage of providing an interface reminiscent to equations in textbooks. However, the executional performance in C/C++ is still tremendous. For example, a similar algebra library forms the basis of the DFT++ project [67]. A significant performance drop arises from the way of how functions return variables. The return variable is typically a local variable of the function (variable "result" in the above examples). This variable is destroyed at the end of the function body. In order to return it to the calling routine, the variable is copied onto the stack and then copied from the stack into the destination variable. This copy operation involves data transactions between RAM and the CPU via the data bus. Note that such data bus operations can be as expensive as the actual numerical operations. When evaluating a vector expression like

$$\mathbf{a} = \alpha(\mathbf{b} + \mathbf{c})\mathbf{d}. \tag{3.22}$$

various temporary objects ($\mathbf{t}_1...\mathbf{t}_3$) will be created

$$\mathbf{a} = \alpha \underbrace{(\mathbf{b} + \mathbf{c})}\mathbf{d} \tag{3.23}$$

$$= \underbrace{\alpha \mathbf{t}_1 \mathbf{d}} \tag{3.24}$$

$$= \underbrace{\alpha \mathbf{t}_2} \tag{3.25}$$

$$= \mathbf{t}_3. \tag{3.26}$$

In BLAS level 1 vector and copy operations require approximately the same executional time. That means that in the above example 7 copy operations (from and to the stack: $2 \times \mathbf{t}_i$ and copy $\mathbf{t}_3 \mapsto \mathbf{a}$) outweigh 3 numerical operations ($\mathbf{a} = \alpha(\mathbf{b} + \mathbf{c})\mathbf{d}$). Due to these additional copy operations functional programming can be significantly slower than procedural programming approaches.

In order to address this issue, in S/PHI/nX the technique of *reference counting* has been adapted to algebraic functions. In this technique an additional counter is attached to every data array. It counts how many variables refer to a vector/matrix. When a vector/matrix is being declared the data array and a counter with the value 1 is created. Every time a variable (or a temporary object) is assigned, the reference counter is increased. Instead of copying the full data array only the pointer is being copied. If an object is being destroyed, the reference counter is decreased. The data array itself is deallocated when the reference counter reaches zero. Instead of returning the full data array only the pointer (8 byte) and the reference counter (4 byte) are copied to and from the stack. In case of the previous example only 56 bytes are copied. By applying reference counting, functional programming becomes as efficient as procedural techniques. These techniques allow to shift memory management, BLAS/LAPACK function call mapping, and data type handling entirely to the compiler.

Computation of Data Types

In the previous paragraphs it has been shown that reference counting and C++ templates allow an efficient *functional* programming approach in contrast to methods applied in most conventional packages which facilitate *procedural* programming. In functional programming all operations can act on temporary objects. Consider a vector expression Eq. (3.22). For every temporary object t_i the compiler needs to know the data type in order to perform the numerical operation. This is, however, not trivial when using C++ templates. In case of many matrix expressions the optimal data type of the resulting temporary object depends also on the *matrix type* and cannot be expressed simply with a template type <T>. The problem can be illustrated, e.g., by means of the efficient computation of eigensystems which are crucial in many parts of S/PHI/nX (e.g., tight-binding initialization, Löwdin orthogonalization). The eigenvalues of a general complex matrix **M** are complex while those of a hermitian matrix **H** are real, i.e.,

$$\mathbf{Mm} = m\mathbf{M} \quad m\epsilon\mathbb{C}$$
$$\mathbf{Hh} = h\mathbf{h} \quad h\epsilon\mathbb{R}, \quad H_{ij} = H_{ji}^*.$$

If this is not considered in the library, all subsequent operations on the eigenvalues would waste CPU time and memory[17]. Therefore, a technique to *"compute"* the computationally optimal data type of temporary algebraic expressions has been developed and implemented in S/PHI/nX. The basis of this approach are "S/PHI/nX type mappers" which create relationships between the various data types. For every type (float, double, single complex, double complex) relations to their corresponding real and complex counterparts (with the same precision) are defined. Any S/PHI/nX data type T can be linked to the matching real or complex type. For example, the type mapper expression "T::Real" refers to the real value of the T. If T is already a real type, "T::Real" is replaced with T. The same applies for "T::Complex". If T is a float "T::Complex" is replaced with the single precision complex type at compile time. All functions in the S/PHI/nX algebra library use such type mapper definitions consistently.

The function to compute eigenvalues of general matrices returns the S/PHI/nX type mapper "T::Complex". If the general matrix was declared as matrix of double values, the resulting eigenvalues are automatically double complex. The function to compute eigenvalues of symmetric/hermitian matrices on the other hand returns "T::Real".

The technique of S/PHI/nX type mappers makes sure that even for temporary expressions, the computationally optimal data type is always being used for the evaluation. For example, the code[18]

```
double a = 10.;
```

---

[17] For example, the multiplication of complex values is computationally 4 times more demanding then the multiplication of real values.
[18] The C++ statement "cout << a << endl" prints the value of a.

```
SxMatrix<Complex16> M = ...;
SxSymMatrix<Complex16> H = ...;
cout << sizeof (a * M.eigenvalues()(0)) << endl;
cout << sizeof (a * H.eigenvalues()(0)) << endl;
```

prints the size (in bytes) of the temporary expressions of the first scaled eigenvalue of a general matrix M and a hermitian matrix H, respectively. The first one is 16 bytes since eigenvalues of a generalized matrix are complex while the second one returns automatically 8 byte, i.e., the size of a double.

S/PHI/nX *computes* the required precision (single / double) of temporary entities. Complex values are applied only if necessary, otherwise the computationally less demanding real types (float, double) will be used. Hence, during the evaluation of algebraic expressions memory consumption can be saved and the number of performed floating point operations can be reduced to a minimum (e.g., number of operations for multiplications). This is crucial when considering algebraic expressions applied to large vectors/matrices.

Automatic error detection

The previously introduced techniques which have been applied/developed to provide an intuitive algebra interface simplify the development process drastically. A major time in the development process is, however, being spent in the debugging process. Typically, a significant amount of time is required in identifying typical errors related to memory management, usage of inconsistent variables, or array index boundary violations. The S/PHI/nX project aims at defining an environment for efficient development of numeric and physical algorithms. Therefore, it is necessary to address the issue of an automatic identification of typical errors related to numerics and physics.

The previously introduced restricted pointers in Fortran allow the vendors of (some) Fortran compilers to provide a very important feature for developers, namely the array boundary checks. Whenever a restricted pointer exceeds array boundaries, an error is being emitted. Such boundary violations typically occur during the process of code development. This Fortran compiler feature can simplify the step of debugging significantly. The typical behavior of the compiler's run-time environment is to cause a program stop when a boundary check was failing. Some compilers also print the code line in which the error occurred. With such an approach only the function in which the violation occurred can be identified. However, the information which is mainly required is the location of the calling procedure instead. Assuming a Fortran pseudo code which provides a function "**trace**" that computes a trace of a matrix:

```
Matrix :: M (n:n)
m = n + 10
a = trace (M, m)      ! ERROR: m exceeds matrix dimensions
subroutine trace (M, n)
   tr = 0
   do i=1, n
      tr = tr + M(i,i)  ! program stops here
   done
   return tr
```

Fortran's array boundary check would identify the boundary inconsistency in the **trace** subroutine. The required information is, however, not the source code line where the error occurred but where the **trace** function was called from. For many applications the execution of a program inside a debugger is often not an option as the execution is too slow for many realistic cases. The time spent for the location of similar problems exceeds often the time of the actual implementation of an algorithm.

In C/C++ the more flexible approach to pointers exclude such an array boundary check. In C/C++ boundary violation can lead to program flaws which might be very difficult to locate. Consider for example the example code

```
int i = 1;
int a[2];
a[2] = 10;              // ERROR: boundary violation!
cout << ''i = '' << i << endl;  // yields: ''i = 10''!
```

The variables will be defined in reverse order on the stack, i.e., `a[0]`, `a[1]`, `i`. Therefore, the memory address of `i` is equivalent to that of `a[2]`. By executing this example code the value of variable `i` is overwritten[19] by the write access of `a[2]`!

Before developing a complex simulation code in C++ a mechanism has to be found to exclude such severe problems. Many huge software projects (such as the Linux kernel[20], the Visualization Tool Kit[21], or KDE[22]) provide boundary checks in many functions. Typically, when a violation is detected a warning message is being printed and the program continues. We consider this standard approach as not suitable for the development of a simulation code.

In S/PHI/nX we introduced an unconventional approach to address this problem. Similar to other C/C++ software projects also in S/PHI/nX functions provide CHECK macros which verify boundaries[23]. Instead of printing warning messages a memory fault is being initiated *intentionally*, namely

```
#define SX_CHECK (expr) if (!expr) { printSomeMessage(); *0 = 1; }
```

Every function in the S/PHI/nX library first verifies boundary accesses using such macros. If a violation is detected the expression "`*0 = 1`" is being evaluated. It tries to assign a value to the pointer to the `NULL` address. This address is, however, located outside of the program segment and, therefore, protected by the operating system. As a result the operating system initiates a segmentation fault and dumps a core file. The generated core file contains all information (variables, stack, thread data) about the process at the time the program stopped. In this case, the core file contains the data of the program when a function was being called with inappropriate arguments. In contrast to Fortran's boundary checks our solution allows an analysis of a core file. A debugger can be used to backtrace the stack after the program execution and the calling function (which caused the misusage) can be easily identified. It is worth mentioning that this approach is significantly faster than executing the program directly in the debugger[24]. Therefore, this method works even for larger test systems.

---

[19]This effect is known as buffer overflow and heavily exploited in various computer hacking techniques.
[20]http://www.kernel.org
[21]http://www.vtk.org
[22]http://www.kde.org
[23]All CHECK macros are activated only in the debug mode. When S/PHI/nX is compiled in the release mode which enables all relevant optimization switches of the compiler, all CHECK macros are ignored.
[24]In order to identify program flaws the code is typically executed directly in the debugger. In this case every function call and memory access is rerouted through the debugger to monitor all stack operations. This slows down the executional performance by a factor of up to 10 (e.g. GNU Debugger gdb 6.6) compared to our approach.

This technique which is the basis of the automatic error detection in S/PHI/nX provides macros beyond mere boundary checks. The library is equipped with macros to verify expressions with respect to divisions by zero (`SX_CHECK_DIV`). Furthermore, macros to verify mathematical identities are provided (`SX_CHECK_VAR`). This macro verifies whether a value is within proper limits. For example, the `sqrt` function that can be element-wise applied to vectors rejects vectors with negative elements[25]. This approach exceeds by far conventional assertion functions since physical identities can be validated. For example, if an algorithm expects an input matrix **H** to be hermitian it would test "`SX_CHECK (H.isHermitian())`". Source code which calls the function with non-hermitian matrices can be instantly identified.

Many Fortran programs do not specify initial values for variables. This can be very dangerous since uninitialized variables depend on the value of the memory at the time the variable is constructed. Many Fortran compilers provide a possibility of initializing all variables with zeros. This is both time consuming and highly unportable as not all Fortran compilers provide this feature. In S/PHI/nX the error detection mechanism has been expanded to track down also uninitialized values without sacrificing executional speed. All floating point variables are initialized with "`sqrt(-1)`" (instead of 0) which is NaN (not a number). All vector/matrix operations are checked for the existence of NaN values. If an algorithm omits the initialization of even a single element S/PHI/nX instantly triggers a segmentation fault and the function in question is identified.

The techniques developed in this project are able to identify the most common development mistakes automatically which improves the speed for development dramatically.

Benchmark results

The above mentioned techniques have been developed and implemented in a general algebraic C++ class library. This algebraic library, which is called SxMath[26], forms the backbone of our *ab-initio* program package. The efficiency[27] of this library is illustrated in Fig. 3.2. For typical vector and matrix sizes commonly used in plane-wave applications performance benchmarks with respect to vector-vector, vector-matrix, and matrix-matrix operations have been performed. The performance that can be obtained with SxMath can significantly outperform existing algebra libraries, such as the BOOST library [113]. The poor performance of Fortran's `matmul` is due to the fact that most compilers do not use BLAS/LAPACK by default. Some compilers, however, support compiler flags to call one of the `_GEMM` BLAS calls. Even if the compilers support it, `matmul` cannot benefit from matrix shapes and matrix packing schemes. The good performance of SxMath in all BLAS levels is due to the consistent combination of above discussed techniques. SxMath also provides an intuitive algebra interface. In Tab. 3.1 some illustrative examples are presented. The syntax is reminiscent to high level toolkits such as Mathematica[28] or Maple[29].

3.2 The Dirac notation in S/PHI/nX

The SxMath library offers a comfortable way to express algebraic equations while guaranteeing high performance. A major challenge of this project was to extend this library such that the C++ compiler "un-

---

[25]This is only an example to demonstrate the concept. In S/PHI/nX a `sqrt` function which returns complex values in this case has been implemented as well.
[26]SxMath: **S/PHI/nX - Math**ematics
[27]The codes have been compiled with `pgf77` and `g++`, respectively.
[28]http://www.wolfram.com
[29]http://www.maplesoft.com

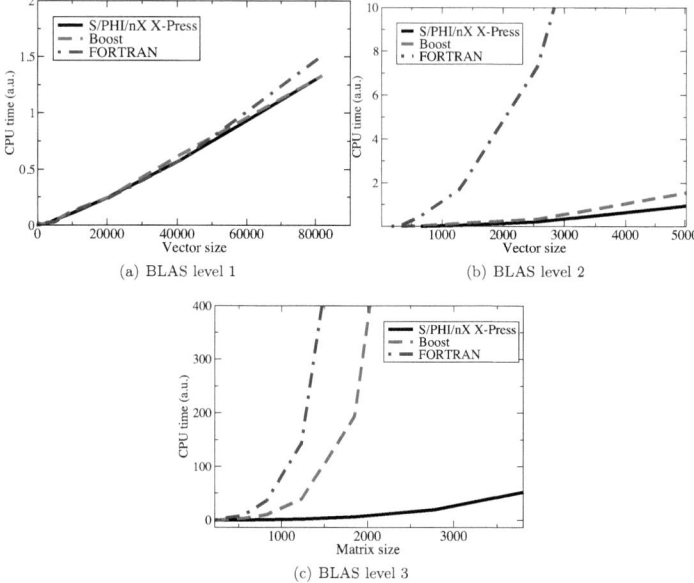

Figure 3.2: Benchmark results of algebra operations using conventional FORTRAN programming (`matmul`), BOOST, and the SxMath library. The vector/matrix sizes have been varied within ranges similar to real plane-wave calculations. A slower slope means better performance. The tests were performed on an Intel P4, 1.6 GHz. The available memory of 512 MB RAM made sure that the tests were not influenced by swapping or paging.
(a) Vector-vector operations: The level caching cannot be used, the main speed-up can be obtained by combining loop unrolling and software-pipelining (see example on page 58). Several operations are performed in parallel on the CPU's pipelines. As test equations of the form $\mathbf{y}=a\mathbf{x}+\mathbf{y}$ have been chosen. Such equations are heavily used in the Gram-Schmidt-orthogonalization, in the update of the wave functions, and the computation of the non-local pseudo potential energy. Note, that for systems containing many atoms (and many states) this routine becomes crucial with respect to executional speed. Both SxMath and BOOST are equally fast, the FORTRAN test performs slightly slower.
(b) Matrix-vector operations: BLAS level 2 allows to use both software-pipelining and level caching. The test system is $\mathbf{y}=\mathbf{Mx}$. Matrix-vector multiplications are applied across the S/PHI/nX package, in particular when updating the gradient $|\Psi\rangle = \hat{H}|\Psi\rangle$. FORTRAN's `matmul` has a poor performance. The BOOST library is almost twice slower than SxMath.
(c) Matrix-Matrix operations: The major speed-up in BLAS level 3 can be accomplished mainly by level caching. A typical candidate to benchmark matrix-matrix operations is the matrix multiplication $\mathbf{A}=\mathbf{MC}$ as used in the tight-binding initialization and the all-state-conjugate gradient as well as the Löwdin orthogonalization. Hence, this operation becomes crucial for all kind of semiconductor/insulator calculations.

Equation	S/PHI/nX
z=ax+b	z = a*x + b;
$C_{ik} = \sum_j A_{ij} B_{jk}$	C=A^B;
$A_{ij} = M_{ij}^{-1}$	A=M.inverse();
$M = LL^t$	L=M.choleskyDecomposition();
$S^{-1} = LL^t$	L=S.inverse().choleskyDecomposition();

Table 3.1: Example of S/PHI/nX expressions using the SxMath library. We tried to keep the S/PHI/nX source code as close as possible to common mathematical toolkits (such as Mathematica). At compile time the code is *replaced* with the corresponding BLAS/LAPACK routines. Therefore, the evaluation of the highly abstract function calls can be accomplished without any loss in performance compared to an explicit BLAS/LAPACK function call.

derstands" the Dirac notation. The highly abstract level of the Dirac notation separates the numerical and basis-set details from the actual algorithm (see Sec. 3.1.2). In order to implement a quantum mechanical algorithm based on expressions given in the Dirac formulation, the numerical and physical context has to be considered by the developer. In the following sections we derive new techniques which allow the C++ compiler to put Dirac expressions in the proper contexts and hence, to perform some of the tasks of physicists/chemists when developing quantum mechanical program packages. It will be shown that the new techniques are able to generate code which is at least as efficient as conventionally developed program packages while a quantum mechanical interface in a Dirac language can be provided. This is crucial since usually higher abstraction results in performance penalties.

In the following sections we illustrate the concepts by means of key contributions to a pseudo potential plane wave Hamiltonian (Eq. 1.46) such as the Hartree potential v_H (Eq. (1.65)), the local v_{loc}^{ps} (Eq. (1.81)) and non-local pseudo potentials v_{nl}^{ps} (Eq. (1.83)), and the kinetic energy $\hat{T}$ (Eq. (1.53)). The chosen equations are typical for the process of developing quantum mechanical program packages. Although the discussions in the following sections focus on these examples, it will be shown that the concept is applicable to all elements of the Dirac notation.

3.2.1 Conventional approach

Most quantum mechanical program packages follow the procedural approach (see Sec. 3.1.4) not only for describing algebraic expressions but also quantum mechanical expressions. The source code files dedicated to the description of the Hamiltonian consist typically of subroutine calls with a huge amount of arguments. In the introduction of this chapter a representative source code example has been presented (see p. 53). Subroutine calls with more than 20 arguments are typical for such approaches. The arguments are mainly necessary to provide dimensions of the basis sets as well as various prefactors. This approach leads to even less transparent source code if global variables are applied. When implementing new features modifications often require massive changes across the code. The implementation of new and advanced algorithms requires a significant amount of time.

3.2.2 Modular approach

With high level languages such as Fortran90/95 and C++ the relevant information can be kept together in modules (Fortran) or classes (C++). For quantum mechanical expressions prefactors and dimensions can be combined with physical entities such as wave functions, potentials, or charge densities. By overloading functions and operators an object[30] can be handled like an intrinsic data type, such as integer or float. For example, the electronic charge density of a pseudo potential plane-wave code would contain the data (density value of $\varrho(\mathbf{R})$ on the FFT grid) as well as functions to manipulate them (e.g., symmetrization, normalization). Default operators, like "+" or "-", would allow the computation of the sum of two densities as simple as "rho = rho1 + rho2".

In order to approach a functional interface the key variables in quantum mechanics must be modularized. That are the wave functions Ψ, basis-sets $\mathbf{B}$, and electronic charge densities ϱ. In modular programming, functions can be defined that act on the self-defined data types. For example, a FFT function call could be added to the wave function type, i.e.,

```
WaveFunctionG psiG = ...;
WaveFunctionR psiR = psiG.fftForward ();
```

Besides wave functions also electronic charge densities need to be transformed between reciprocal and real space. The same transformation for charge densities would read

```
RhoG rhoG = ...;
RhoR rhoR = rhoG.fftForward ();
```

This example shows that functions (like Fourier transformations) should not be added directly into the wave function or charge density object since redundant code would have to be implemented. A reduction of redundancy in the source code can be accomplished by separating operator functions from data types. Operator functions are related to the basis-sets. Hence, for $\mathbf{R}$, $\mathbf{G}$, $\mathbf{G} + \mathbf{k}$, and r own data types can be defined. They contain dimensions, prefactors arising for transformations, and the projector functions.

In the following it will be discussed how wave functions and basis-sets are modularized in order to allow the compiler a detection of the corresponding physical/numerical contexts.

Bloch-like wave functions

According to Sec. 1.4 Bloch-like wave function coefficients can be represented either in vector $c_{i\sigma\mathbf{k}}(\mathbf{G})$ or in matrix form ($C_{i\mathbf{G}}(\sigma\mathbf{k})$). The relevant quantum numbers to specify a state are the Bloch state i, the spin channel σ. Furthermore, the $\mathbf{k}$ point is necessary to specify the state.

In Sec. 3.1.3 the advantage of blocking matrix algorithms has been discussed. In the matrix representation each Bloch state i is kept as a column with n_{pw} row entries

$$\langle \mathbf{G} + \mathbf{k}|\Psi_{\sigma\mathbf{k}}\rangle_{i\mathbf{G}} = C_{i\mathbf{G}}(\sigma\mathbf{k}) = \begin{pmatrix} C_{11}(\sigma\mathbf{k}) & C_{21}(\sigma\mathbf{k}) & \cdots & C_{n1}(\sigma\mathbf{k}) \\ C_{12}(\sigma\mathbf{k}) & C_{22}(\sigma\mathbf{k}) & \cdots & C_{n2}(\sigma\mathbf{k}) \\ \vdots & \vdots & \ddots & \vdots \\ C_{1n_{\mathrm{pw}}}(\sigma\mathbf{k}) & C_{2n_{\mathrm{pw}}}(\sigma\mathbf{k}) & \cdots & C_{n_{\mathrm{pw}}}(\sigma\mathbf{k}) \end{pmatrix}. \tag{3.27}$$

---

[30] An object is an instance of a class.

Property	Value		Interface
Construction	$\|\tilde{\Psi}_{\mathbf{G+k}}\rangle$	$\sum_{\mathbf{G}} C_{i\sigma\mathbf{k}}(\mathbf{G})e^{i\mathbf{G}\cdot\mathbf{r}}$	SxPW psi;
Referencing / Quantum numbers			
Single state	$\|\tilde{\Psi}_{i\sigma\mathbf{k}}\rangle$	Vector: $c_{i\sigma\mathbf{k}}(\mathbf{G})$	psi(i,iSpin,ik);
All states	$\|\tilde{\Psi}_{\sigma\mathbf{k}}\rangle$	Matrix: $C_{i\mathbf{G}}(\sigma\mathbf{k})$	psi(iSpin,ik);

Table 3.2: Summary of the key functions to support Dirac notation for Bloch waves. Bloch wave functions are implemented as containers of algebraic vectors. By choosing one of the referencing operators, either (i,iSpin,ik) or (iSpin,ik), one comfortably can switch between vector or matrix representation.

In S/PHI/nX all states are stored in a SxMath object of type SxMatrix<Complex16>. By exploiting the reference counting technique (see page 63) the full matrix is returned at once using the intuitive interface psi(iSpin,ik). Subsequent operations on the object psi(iSpin,ik) can provide full BLAS3 support.

As mentioned previously (e.g., page 56) beside the computationally efficient matrix form of wave functions, a vector representation is required to reduce the memory consumption. The vector representation considers the ith column vector of the previous matrix individually

$$\langle \mathbf{G} + \mathbf{k}|\Psi_{i\sigma\mathbf{k}}\rangle_{\mathbf{G}} = C_i(\sigma\mathbf{k}) = \begin{pmatrix} C_{i1}(\sigma\mathbf{k}) \\ C_{i2}(\sigma\mathbf{k}) \\ \vdots \\ C_{iN_{\mathrm{pw}}}(\sigma\mathbf{k}) \end{pmatrix}. \qquad (3.28)$$

The interface psi(i,iSpin,ik) returns a single column of the matrix $C_{i\mathbf{G}}(\sigma\mathbf{k})$. In the approach of Ref. [67] this operation is accomplished by copying an entire column which is computationally very demanding. In S/PHI/nX, the reference counting technique has been adapted such that only a pointer to the first column element is returned and the memory management of the returned vector object is deactivated. Instead, the matrix handles the memory management. This approach has the advantage, that wave functions can simultaneously be accessed in both matrix and vector shape. Since only pointers (8 bytes) are copied all operations acting on wave functions are very efficient. The S/PHI/nX interface for Bloch-like wave functions is shown in Tab. 3.2.

The return value of both psi(i,iSpin,ik) and psi(iSpin,ik) is a vector or a matrix of the above discussed high-performance algebra library. Therefore, algebraic expressions on wave function coefficients will be automatically optimized. For example, the code fragment to compute an overlap matrix

$$\mathbf{S}_{ij} = \langle \Psi_{i\sigma\mathbf{k}}|\Psi_{i\sigma\mathbf{k}}\rangle \qquad (3.29)$$

would read

```
SxPW psi = ...;
int iSpin = 0, ik = 0;
SxSymMat<Complex16> S(nStates);
for (int i=0; i < nStates; ++i)
   for (int j=i; j < nStates; ++j)
      S(i,j) = psi(i,iSpin,ik) ^ psi(j,iSpin,ik);
```

Property	Value	Interface
Construction	$|\mu_r\rangle$	SxAtomicOrbitals mu;
Referencing:		
Single state	$|\mu_{i_s i_a nlm\sigma \mathbf{k}}\rangle$	$\hat{T}_{i_s i_a}|\phi_{i_s nlm\sigma}\rangle$	mu(iSpecies,iAtom,n,l,m,iSpin,ik);
Index map	$|\mu_{\tilde{i}\sigma \mathbf{k}}\rangle$	$|\mu_{\tilde{i}\mapsto i_s i_a nlm}\rangle$	mu(i,iSpin,ik);

Table 3.3: Summary of the key functions to support Dirac notation for atomic orbitals.

and is replaced with ZDOTC BLAS calls during the compilation. The class representing the Bloch-like wave functions is only a *container* of algebraic vectors. The actual mathematical calculation is performed in BLAS calls mapped by the SxMath library.

Atomic orbitals

In a pseudo potential approach atomic orbitals are used to represent the pseudo wave functions. Atomic orbitals $|\mu\rangle$ are characterized by the quantum numbers n, l, m as well as the species and atomic indices i_s and i_a. Atomic orbitals in S/PHI/nX provide two index maps. The map $\tilde{i}_{\text{ref}} \mapsto (i_s nlm)$ refers to a reference orbital, i.e., $|\phi_{i_s nlm}\rangle$, or in S/PHI/nX syntax mu(is,n,l,m). The second index map $\tilde{i} \mapsto (i_s i_a nlm)$ refers to an orbital shifted to the location of atom i_a. By providing the additional parameter the corresponding S/PHI/nX interface is simply mu(is,ia,n,lm).

This second interface to address an atomic orbital is equivalent to an application of the translation operator $\hat{T}_{i_s i_a}$

$$|\mu_{i_s i_a nlm\sigma}\rangle = \hat{T}_{i_s i_a}|\phi_{i_s nlm\sigma}\rangle. \tag{3.30}$$

Beside the internal management of reference and atomic orbitals the two index arrays can simplify the source code when applying operations f to all orbitals *sequentially*. Using the index map $\tilde{i}$ the code

```
for (int is=...)
   for (int ia=...)
      for (int n=...)
         for (int l=...)
            for (int m=...)
               muNew = f (mu(is,ia,n,l,m));
```

can be transformed to a single loop over all orbitals

```
for (int iOrb=0; iOrb < mu.getNOrbitals(); ++iOrb)
   muNew = f (mu(iOrb));
```

In Tab. 3.3 the S/PHI/nX interface for atomic orbitals is defined.

New wave functions

The above mentioned interfaces to represent wave functions can easily be extended in future code developments. Let us assume an extension of S/PHI/nX to support PAW. In this case the wave functions need to be represented as Bloch-like waves as well as orbital-like partial waves. The modular concept would support such an extension as follows.

The all-electron wave function consists of a smooth auxiliary wave function $|\tilde{\Psi}\rangle$ as well as the partial waves $|\tilde{\phi}\rangle$ and $|\phi\rangle$ (SxAtomicOrbital) (see Sec. 1.5.2). The new entity to represent wave functions in PAW would refer to SxPW and SxAtomicOrbital in the referencing operator. In order to provide an intuitive interface operators like "+" and "-" can be overloaded.

Basis-sets

So far the modularized form of the wave function class connects an intuitive interface with high performance algebra function calls. However, when developing a class describing the Hamiltonian various quantum mechanical operations (e.g. Eq. (1.64), Eq. (1.53), Eq. (1.63)) are applied to wave functions. A mere algebraic interface is not sufficient here since these operations depend on the choice of the basis-set. This work aims for deriving methods that allow an Hamiltonian to be implemented independently of the basis-set. Hence, the basis-set dependent operators should be modularized in terms of basis-set classes.

In a pseudo potential plane-wave approach, Bloch-like wave functions are represented in real space $\langle \mathbf{R}|$ (application of the effective potential v_{eff}, Eq. (1.24)) and reciprocal space $\langle \mathbf{G}+\mathbf{k}|$ (kinetic energy and non-local potential, T Eq. (1.53) and v_{nl}, Eq (1.83)). Basis-set dependent operators are the evaluation of the trace (Eq. (1.51)), the Laplacian ∇^2, and the definition of a metric for scalar products. The radial space $\langle r|$ is necessary to describe atomic orbitals. Furthermore, projections between basis-sets have to be defined in the same scope. Such an approach is also flexible with respect to a future extension to e.g. PAW. The smooth auxiliary wave function $|\tilde{\Psi}\rangle$ is sampled on $\langle \mathbf{G}+\mathbf{k}|$ and $\langle \mathbf{R}|$ while the partial waves $|\phi\rangle$ and $|\tilde{\phi}\rangle$ are projected on $\langle r|$.

In Tabs. 3.4, 3.5, and 3.6 the S/PHI/nX interfaces to $\langle \mathbf{R}|$, $\langle \mathbf{G}+\mathbf{k}|$, and $\langle r|$ are presented.

Hamiltonian

The modularized wave function and basis-set approach extends the functional interface which reflects already the key elements of the Dirac notation. This can be illustrated by some examples of the Hamiltonian.

```
SxGBasis G;
SxGkBasis Gk;
SxRBasis R;
SxRhoG rhoG;
SxRhoR rhoR = G.projectTo (R, rhoG);
```

With the above variable declarations the kinetic energy defined in Eq. 1.53 becomes

```
for (i=0; i < nStates; ++i)
   psi = waves(i,iSpin,ik)
   for (iSpin=0; iSpin < nSpin; ++iSpin)
      for (ik=0; ik < nk; ++ik)
         eKin += omega(k) * focc(i,iSpin,ik) * Gk(ik).laplacian(psi,psi);
```

Property	Value	Interface
Sampling:		
Grid points	$\mathbf{R}_{ijk}$	$\frac{i}{N_1}\mathbf{a}_1 + \frac{j}{N_2}\mathbf{a}_2 + \frac{k}{N_3}\mathbf{a}_3$	SxRBasis R;
Abstract wave function representation:		
Identity	$\|\Psi_\mathbf{R}\rangle$	$\sum_\mathbf{R} \|\mathbf{R}\rangle\langle\mathbf{R}\|\Psi\rangle$	psi;
		or:	(R|psi);
Integration and Metric:		
Scalar product	$\langle a\|b\rangle$	$\sum_{ijk} a^*_{ijk} b_{ijk}$	(a|b);
Trace	$\mathrm{tr}_\mathbf{R} X(\mathbf{R})$	$\Delta\Omega \sum_{ijk} X(\mathbf{R}_{ijk})$ with $\Delta\Omega = \frac{\|\mathbf{a}_1\cdot(\mathbf{a}_2\times\mathbf{a}_3)\|}{(N_1 N_2 N_3)}$	tr(X);
Projectors:		
Project wave functions to G	$\langle \mathbf{G}+\mathbf{k}\|\mathbf{R}\rangle$	$\frac{1}{\sqrt{\Omega}} e^{-i(\mathbf{G}+\mathbf{k})\cdot\mathbf{R}}$	(Gk|R);
Project densities/potentials to G	$\hat{\mathcal{G}}_\mathbf{R}$	$\frac{1}{\sqrt{\Omega}}\sum_\mathbf{R} e^{-i\mathbf{G}\cdot\mathbf{R}}$	X.toG ();

Table 3.4: Summary of the key functions to support Dirac notation in real space contexts. Various Dirac objects (property column) are presented with its physical values as well as the S/PHI/nX programming interface. It can be seen that the interface is strongly reminiscent to the Dirac notation and the resulting source code becomes very intuitive.

while via Eqs. (1.61), (1.62), (1.65), and (1.66) the Hartree potential/energy contribution is:

```
gaussianFunc = ...
for (is = ...)
    rhoGaussG += G.phase(is) * r.projectTo (G, gaussianFunc);   // Eq. (1.62)
vHartree = FOUR_PI/G.gVec(SxIdx(1:ng)) * rhoGaussG;             // Eq. (1.65)
eHartree = 0.5 * G.trace (rho * vHartree);                      // Eq. (1.66)
```

The local pseudo potential/energy contribution Eqs. (1.80), (1.81), (1.82) can be expressed as:

```
locPsFunc = ...
for (is = ...)
    vLocG += G.phase(is) * r.projectTo (G, locPsFunc);          // Eq. (1.81)
eLocPs = R.trace (rhoR * G.projectTo (R, vLocG));               // Eq. (1.82)
```

The contributions to the gradient (Eqs. (2.5), (2.5)) can also be transcribed easily:

```
PsiG psiG;
PsiR psiR = Gk(ik).projectTo (R, psiG);
dPsiG  = 0.5 * Gk(ik).g2 * psiG;                                // Eq. (2.5)
dPsiG += Gk(ik).projectTo (Gk(ik), vEffR(iSpin) * psiR);        // Eq. (2.7)
```

Due to the modularization of quantum mechanical key entities in wave functions and basis-sets a Hamiltonian can be implemented very intuitively. The transcription of expressions given in the Dirac notation into

Property		Value		Interface
Sampling:				
Grid **G**	$\mathbf{G}_{ijk}$	$\mathbf{b}_1 i + \mathbf{b}_2 j + \mathbf{b}_3 k$		SxGBasis G;
Grid $\mathbf{G}+\mathbf{k}$ (full mesh)	$(\mathbf{G}+\mathbf{k})_{ijk}$	with $\mathbf{b}_i = 2\pi \frac{b_j \times b_k}{b_i \cdot b_j \times b_k}$ and cyclic		SxGkBasis Gk;
condensed mesh	$\mathbf{G}_{ig}$ $(\mathbf{G}+\mathbf{k})_{ig}$	$0 \leq \|\mathbf{G}_{ijk}\|^2 \leq G_{\text{cut}}$ $0 \leq \|(\mathbf{G}+\mathbf{k})_{ijk}\|^2 \leq E_{\text{cut}}$		G.gVec(ig); Gk.gVec(igk);
Abstract wave function representation:				
Identity	$\langle\mathbf{G}+\mathbf{k}\|\Psi_\mathbf{k}\rangle$	$\sum_{\mathbf{G}+\mathbf{k}} \|\mathbf{G}+\mathbf{k}\rangle\langle\mathbf{G}+\mathbf{k}\|\Psi_\mathbf{k}\rangle$		psi;
			or:	(Gk|psi);
Integration/Metric:				
trace	$\text{tr}(\hat{\varrho}\hat{A})$	$\sum_\mathbf{G} \varrho(\mathbf{G})\hat{A}^*(\mathbf{G})$		tr(rho*A);
Operators:				
Laplacian	$\langle\mathbf{G}+\mathbf{k}\|\hat{L}\|\tilde{\Psi}_{i\sigma\mathbf{k}}\rangle$ $\hat{L}C_{i\mathbf{G}}(\sigma\mathbf{k})$	$\|\mathbf{G}+\mathbf{k}\|^2\|c_{i\sigma\mathbf{k}}(\mathbf{G})\|^2$ $(\|\mathbf{G}+\mathbf{k}\|^2 C_{i\mathbf{G}}(\sigma\mathbf{k}))$		waves(i,s,k).laplacian(); waves(s,k).laplacian();
Translator	$\langle\mathbf{G}\|\hat{T}_{i_s}\|\Phi_{i_s}\rangle$	$\mathbf{S}_{i_s}(\mathbf{G})\Phi_{i_s}(\mathbf{G})$		G.T(is)*phi;
Translator	$\langle\mathbf{G}+\mathbf{k}\|\hat{T}\|\mu\rangle$			
Projectors:				
to realspace	$\langle\mathbf{R}\|\mathbf{G}\rangle$ $\hat{\mathcal{R}}_\mathbf{G}$	$\Delta\Omega \sum_\mathbf{G} e^{+i(\mathbf{G}+\mathbf{k})\cdot\mathbf{R}}$ $\Delta\Omega \sum_\mathbf{G} G e^{+i\mathbf{G}\cdot\mathbf{R}}$		(R|G); X.toR();

Table 3.5: Summary of the key functions to support Dirac notation for G contexts. Sampling and integration are performed on the regular (FFT) grid, the main projections are to R space using FFT.

Property	Value		Interface
Sampling:			
Grid points	r_i	$r_i = r_{i+1}/\Delta r_{\log}$	SxRadBasis r;
Wave functions	$\|\mu_r\rangle$ $\langle r\|\mu\rangle$		SxAtomicOrbitals mu(r);
Projectors:			
Identity	$\langle r\|r\rangle$	1	(r|r);
Project wave functions to R	$\hat{\mathcal{G}}_r$	$\sqrt{\frac{2l+1}{4\pi}} \frac{4\pi}{\sqrt{\Omega}} \int_0^\infty dr\, r^2 j_l(\|\mathbf{G}\|r)$	(G|r);

Table 3.6: Summary of the key functions to support Dirac notation for r contexts. Sampling and integration are performed on an logarithmic mesh, the main projections are to **G** space.

source code can be accomplished in significantly shorter time than conventional procedural functions. Since equations can be implemented almost directly from the Dirac notation we call this approach quasi-Dirac notation.

The quasi-Dirac notation requires that the wave functions and basis-sets have to be explicitly provided. This is error-prone since it must be ensured by the developer that basis-set and wave function objects always match. This requirement makes an cumbersome management of such objects throughout the code necessary. In the following section a way of solving this drawback will be shown.

3.2.3 Object-oriented approach

In order to obtain an intuitive programming meta language for quantum mechanical expressions the previously introduced quasi-Dirac ansatz has to be extended such that the compiler "understands" the Dirac notation. In this section the required object-oriented programming techniques will be derived.

In order to detect the quantum mechanical context (see Sec. 3.1.3) a further degree of modularization is necessary. C++ provides language elements for object oriented programming (OOP). As object orientation is not yet a very common approach in the field of high performance computing its ideas are briefly sketched here. In procedural languages, such as FORTRAN or Fortran90/95[31], a program is a collection of functions and subroutines which build up an algorithm. In OOP, the program is organized in terms of self-defined data types (classes) which contain both data as well as functions which can act on this data. In contrast to modular programming, in OOP hierarchies of classes can be defined. More complex classes can derive properties from basis classes. For example, a density used in PAW would be defined of three contributions, the plane-wave density $\tilde{\varrho}(\mathbf{R})$ as well as the contribution inside the one-center densities $\varrho_i(r)$ and $\tilde{\varrho}_i(r)$. By applying OOP both densities could be defined in separate classes to represent radial densities. An abstract density class could then combine $\varrho(\mathbf{R})$, $\varrho(r)$, and $\tilde{\varrho}(r)$ in a new type for describing PAW densities. By deriving from the basis classes (e.g. SxPW and SxAtomicOrbitals) the new class inherits all functions of the basis classes. The functionalities implemented in the basis classes are available in the derived classes without additional programming lines which simplifies the source code dramatically:

```
class A {
   public:
      void foo1();
};
class B {
   public:
      void foo2();
};
class C : public A, public B  { /* derive from A and B */ };

C obj;
obj.foo1 ();   // call derived function from A
obj.foo2 ();   // call derived function from B
```

---
[31]Fortran90/95 is not object oriented although sometimes claimed otherwise. Fortran90/95 lacks the ability of polymorphism, i.e., the capability of inheriting classes from various basis-classes. Polymorphism is one of the 3 required features of object oriented programming besides inheritance and encapsulation. It will be shown that all 3 features are necessary to support Dirac's notation in the source code.

An important step in accomplishing a meta language that supports Dirac's notation is the abstraction of wave functions and basis-sets. Instead of operating explicitly on coefficient arrays (SxPW, SxAtomicOrbitals) or the basis-sets (SxRBasis, SxGBasis, SxRadBasis), in the Dirac notation more general wave functions $|\Psi\rangle$ or basis-sets $\langle \mathbf{X}|$ are used. Generalization of data types in C++ is conventionally accomplished by specifying *virtual functions* (placeholders) in the basis classes and specifying the actual function in the derived class. This approach can be illustrated by means of a Laplacian:

```
class SxBasisSet {
   public:
      virtual Psi laplacian (Psi);    // empty placeholder
};
class SxPW {
   public:
      virtual Psi laplacian (Psi psiIn) {
         return Gk.laplacian (psiIn);  // perform laplacian in <G+k| basis
      }
};
class SxAtomicOrbitals {
   public:
      virtual Psi laplacian (Psi psiIn) {
         return rad.laplacian (psiIn); // performs laplacian in <r| basis
      }
};
```

The advantage of such a construction is that an identical interface can be applied for both wave function representations:

```
SxPW psiG = ...;
SxAtomicOrbital mu = ...;
lPsi = psiG.basis->laplacian (psiG);
lMu  = mu.basis->laplacian (mu);
```

This identical interface for two very different functions can be only accomplished by attaching the information about the basis to each wave function. Here a crucial compromise has to be made:

The actual numerical calculations cannot be performed with an abstract Dirac vector $|\Psi\rangle$. Instead, a vector is sampled on the basis-set and it is represented as expansion coefficients $\langle \mathbf{B}|\Psi\rangle = c(\mathbf{B})$. In conventional codes the algorithms act directly on these expansion coefficients. As a result those codes are strongly basis-set dependent. In order to create a quantum physics library which can offer an interface using *abstract* Dirac vector objects every $|\Psi\rangle$ needs to "know" the basis-set on which it is sampled on. In order to bridge the gap between the notation of the abstract Dirac vector $|\Psi\rangle$ and the computationally necessary basis-set dependent coefficient notation $c(\mathbf{B}) = \langle \mathbf{B}|\Psi\rangle$ we introduce a new symbol $|\Psi_\mathbf{B}\rangle$. The index $\mathbf{B}$ symbolizes that for an introduction of a Dirac vector in the source code, the basis-set has to be provided. This needs to be done once, namely when constructing the vector. For the entire lifetime of the vector it keeps the information about its basis-set $\mathbf{B}$. For all operations acting on the vector, $\mathbf{B}$ does not need to be provided anymore by the developer. All algorithms acting on the new object $|\Psi_B\rangle$ can be now formulated *independently* of the chosen basis-set.

The compromise $|\Psi_B\rangle$ has been implemented like the following example:

```
class SxPsi {
   public:
      SxBasis *basis;            // pointer to ''any'' basis
};
class SxPW : public SxPsi {
   public:
      SxPW (SxBasis *basis);     // specify basis pointer
};
class SxAtomicOrbitals : public SxPsi {
   public:
      SxAtomicOrbitals (SxBasis *basis); // specify basis pointer

};
// --- (A) Constructing Dirac elements.
SxGkBasis Gk = ...;
SxPW psiG (&Gk);              // <G+k|psi>
SxAtomicOrbitals mu (&rad);   // <rad|mu>

// --- (B) Transition from coefficient vectors to abstract Dirac vectors.
SxPsi psi = psiG;
// OR:
SxPsi psi = mu;

// --- (C) Type of psi and basis is ''hidden'' from here on.
//         Using Dirac notation, no coefficient vectors anymore.
SxPsi dPsi = psi.basis->laplacian ();
```

During the construction of wave functions (A) the information about the basis-set is attached to the object. Here, a wave function is represented as coefficient array $c(\mathbf{B})$ sampled on a specified basis $\mathbf{B}$. In the source code block (B) the specified wave function type is converted to an unspecified wave function (wave function basis class SxPsi), i.e., $c(\mathbf{B}) \rightarrow |\Psi_B\rangle$. In block (C) an operator (Laplacian $\hat{L}_B$) can operate on an object symbolizing a wave function $|\Psi\rangle$ instead of a coefficient array. A Hamiltonian can be implemented using only the abstract wave functions $|\Psi_B\rangle$ as well as operators defined in the basis-set $\langle\mathbf{B}|$. The corresponding part of the S/PHI/nX class hierarchy is shown in Tab. 3.3. In Tab. 3.7 all elements of the S/PHI/nX Dirac notations are presented.

In quantum mechanical expressions the introduction of identities is often useful. In order to make sure that such identities do not generate machine code (which would slow down the performance) the S/PHI/nX Dirac

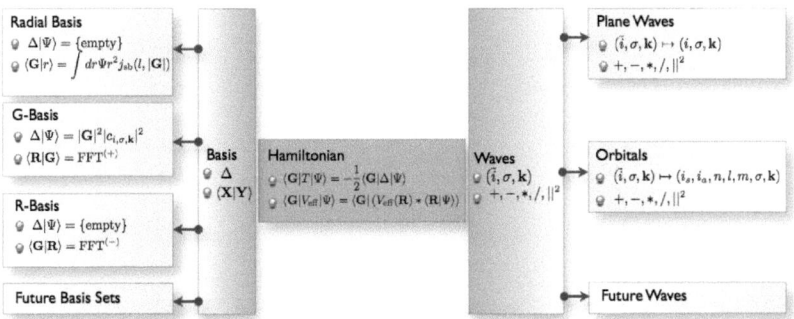

Figure 3.3: Illustration of the various abstraction levels in the S/PHI/nX class hierarchy by means of the Hamiltonian: (I) In the *source code* of the Hamiltonian (central box) a writing style reminiscent to the Dirac notation is realized using '(', '|', and ')' to build bras and kets. Additionally operators like the Laplacian are used in the Hamiltonian. (II) *At compile time* all those operators are replaced with function calls in the abstract basis-set interface (box left of the Hamiltonian). Depending on which basis-set is used the actual implementation of the operators (projectors, Laplacian, etc.) are invoked. (III) The same approach is used in the case of wave functions: The Hamiltonian interacts with an abstract wave function class (box right of the Hamiltonian) to extract a single state. Depending on the used wave functions the actual wave function container is used instead.

basis-set independent bra-ket notation	basis-set dependent bra-ket notation	formal S/PHI/nX interface	internal S/PHI/nX representation
$\|\Psi\rangle$	$\sum_\mathbf{B} \|\mathbf{B}\rangle\langle\mathbf{B}\|\Psi\rangle$	$\|\Psi_\mathbf{B}\rangle$	$\langle\mathbf{B}\|\Psi\rangle = c(\mathbf{B})$
$\langle\mathbf{B}\|\Psi\rangle$	$\langle\mathbf{B}\|\sum_\mathbf{B}\|\mathbf{B}\rangle\langle\mathbf{B}\|\Psi\rangle$	$\langle\mathbf{B}\|\Psi_\mathbf{B}\rangle$	$\langle\mathbf{B}\|\Psi\rangle = c(\mathbf{B})$
$\langle\mathbf{X}\|\Psi\rangle$	$\langle\mathbf{X}\|\sum_\mathbf{B}\|\mathbf{B}\rangle\langle\mathbf{B}\|\Psi\rangle$	$\langle\mathbf{X}\|\Psi_\mathbf{B}\rangle$	$\sum_\mathbf{B}\langle\mathbf{X}\|\mathbf{B}\rangle\langle\mathbf{B}\|\Psi\rangle = c'(\mathbf{X})$
$\hat{A}$		$\hat{A}_\mathbf{B}$	$\langle\mathbf{B}\|\hat{A}\|\mathbf{B}\rangle$
$\hat{O}$		$\hat{O}_\mathbf{B}$	$\langle\mathbf{B}\|\hat{O}\|\mathbf{B}\rangle$
$\hat{L}$		$\hat{L}_\mathbf{B}$	$\langle\mathbf{B}\|\hat{L}\|\mathbf{B}\rangle$
$\mathrm{tr}(\hat{\varrho}\hat{A})$		$\mathrm{tr}_\mathbf{B}$	$\mathrm{tr}_\mathbf{B}(\hat{\varrho}\hat{A}_\mathbf{B})$

Table 3.7: Comparison between Dirac's notation (1st and 2nd column) and our basis-set "aware" S/PHI/nX notation (3rd and 4th column). Beside the wave functions also operators ($\hat{A}$) such as the overlap $\hat{O}$ or Laplacian operators $\hat{L}$ need to have access to the basis-set. Also the evaluation of a trace is basis-set dependent.

Expression	Template
$\langle G\|R\rangle$	`return SxProjector<SxRBasis,SxRBasis> (&a,&b);`
$\langle R\|G\rangle$	`return SxProjector<SxRBasis,SxGBasis> (&a,&b);`
$\langle r\|G\rangle$	`return SxProjector<SxRadBasis,SxGBasis> (&a,&b);`
$\langle G\|r\rangle$	`return SxProjector<SxRBasis,SxGBasis> (&a,&b);`

Table 3.8: Registration of bra-ket combinations in S/PHI/nX. An extension to new basis-sets is trivial.

notation has to be extended with a basis-set dependent identity[32]

$$|1\rangle\langle 1| \Rightarrow |1_B\rangle\langle 1_B|. \tag{3.31}$$

Applying these "new" basis-set aware operators and wave functions, the Hamiltonian and its application on wave functions can be accomplished entirely basis-set independently. The same technique which has been applied to generalize wave function classes can be adapted to transformations:

```
psiG.basis->projectTo (R);    // SUM_G <R|G><G|psi> or simply <R|psi>
```

With the generalization classes SxPsi and SxBasis the Dirac notation employing *bra* and *ket* vectors can eventually be defined. Therefore, a template class (see p. 62) to represent the *bra-ket* elements

```
template<class Bra, class Ket>
class SxProjector {
   public:
      Bra *braPtr;   // pointer to <bra|
      Ket *ketPtr;   // pointer to |ket>
};
```

can be defined. In order to provide a full Dirac notation the "|" operator[33] has been overloaded according to all possible combinations (see Tab. 3.8). During the compilation an expression like "(G|R)" will be *replaced* with the proper FFT function call. Please note that this source code expression returns a function rather than a value! The interface is strongly reminiscent to the expression in the Dirac notation $\langle G|R\rangle$ which allows a straightforward implementation of various quantum mechanical projectors without any loss in computational performance!

Similar Dirac expressions have to be supported for wave functions such as $\langle X|\Psi\rangle$. Since the numerical operation depends in this case on the context (see e.g. Eqs. (3.18)-(3.21)) a different solution has to be found:

```
template<class T>
operator| (SxRadBasis b, SxDiracVec<T> v)
{
```

---
[32] This identity operator is not necessarily required. In reality it is used in particular to verify that the basis-sets left and right of the identity do match. Without such test facilities the high abstraction level could easily lead to barely traceable program flaws.
[33] The default behavior of the "|" operator in C is the evaluation of the bit-wise OR operation.

```
    return v.basis->projectTo (&b, &v);
}
template<class T>
operator| (SxDiracVec<T> v, SxRadBasis b)
{
    return (b | v).conj());
}
```

"Virtual" templates In principle, this technique could be applied to all other basis-sets. Unfortunately, this approach as presented so far is in violation to C++ since the virtual projector functions (`projectTo`) contain template arguments (`SxDiracVec<T> v`). One of the fundamental language concepts in C++ is type-safety. Virtual functions can be discriminated only by their function arguments. If these arguments are templates, distinguishing virtual functions becomes impossible for the compiler. Therefore, the usage of templates and virtual functions are mutually excluded in C++. In our ansatz, however, the usage of virtual functions is crucial to create abstract wave function (p. 77) and basis-set objects (p. 78). Also template arguments are substantial in our approach to obtain optimal numerical performance (pp. 62). In order to address this problem the required C++ type safety has to be disabled temporarily. Therefore, in S/PHI/nX the actual type of vector v is removed and replaced with a void pointer[34]. The above function call becomes

```
    SxDiracVec<T> vec = ...;
    void *vecPtr = (void *)(&v);        // type cast to void
    projectTo (b.getBasisPtr(), vecPtr); // Dirac vector has no type
```

Due to this type cast all function arguments are well defined and `projectTo` can be a virtual function that is defined in the basis class **SxBasis** (p. 78). The pointer to the corresponding object can be extracted from b. An abstraction of wave functions and basis-set is now possible. When performing the actual projector operation the vector type is, however, required. For example, when performing a FFT function call it is important whether the FFT mesh describes real or complex values. The above introduced type cast removes all type information completely. Algebraic vectors can only contain real or complex values with either single or double precision. In a first attempt the information about which of these types have to be considered in the projector has been "attached" by casting any number (e.g. 0) to one of these types using S/PHI/nX type mappers (p. 64):

```
    projectTo (b.getBasisPtr(), vecPtr, (b::Type)0, (v::Type)0);
```

This solution still did not solve the actual problem, but the discrimination of input and output data type of the Dirac projector function is now simpler to handle. With the two additional arguments one out of 16 `projectTo` functions[35] can be easily be selected.

```
    #define VIRTUAL_PROJECT_TO (B) \
       projectTo (B *, void *, double, SxComplex16); // vector is C16, returns double
       ...
       projectTo (B *, void *, SxComplex16, double); // vector is double, returns C16
       ...
```

---
[34]The C/C++ data type void refers to *no type*.
[35]Permutation of all 4 floating point types)

The necessary 16 functions can be joined in a single preparser text macro which will be added to every basis-set class. The actual projector can then overload these functions in the derived class. Therefore, the proper projector can be registered using the following macro

```
#define REGISTER_PROJECTOR(FROM,TO,PROJECTOR) \
   virtual void projectTo (const TO, const void *in, void *o, float, double) {...}
   ...
   virtual void projectTo (const TO, const void *in, void *o, float, SxComplex16) { ... }
   ...
```

With this set of macros a combination of virtual functions with (a limited number of) template arguments can be accomplished within the C++ language concepts. The macros are available in S/PHI/nX service files and do not need to be modified for future code developments. For each projector every basis-class needs exactly one source code line. For example, the class for representing the **G**-basis, the identity $\mathbf{1_G}$ (Eq. (3.31)) and $\langle \mathbf{R}|\mathbf{G}\rangle$ (Eq. (1.48)) need to be registered:

```
class SxGBasis : public SxBasis
{
   ...
   REGISTER_PROJECTOR (SxGBasis, SxGBasis, identity);  // (G | psiG) = psiG
   REGISTER_PROJECTOR (SxGBasis, SxRBasis, toRSpace);  // (R | psiG) = FFT (psiG,psiR)
   ...
};
```

With these two additional lines the source code "(G | psiG)" will be replaced with the function call "identity" while "(R | psiG)" will perform a FFT function call.

By applying the predefined S/PHI/nX macros, registering new basis-sets into the existing Dirac environment of S/PHI/nX can be accomplished by copying these few source lines (REGISTER_PROJECTOR). Therefore, introduction of new Dirac projectors in S/PHI/nX is extremely simple.

Dirac's *bra-ket* notation is appropriate when describing wave functions. Projections of entities like charge densities or potentials cannot be symbolized in the *bra-ket* style. In order to project those terms as simple as wave functions we introduce new basis-set operators written in calligraph letters. An entity given in any basis-set **X** can be projected onto the basis-set **B** with the operator $\hat{\mathcal{B}}_{\mathbf{X}}$. For example, a density in basis-set **B** can be projected to real space **R** like

$$\varrho(\mathbf{R}) = \hat{\mathcal{R}}_{\mathbf{G}}\varrho(\mathbf{G}) \qquad (3.32)$$

or back to **G** space with

$$\varrho(\mathbf{G}) = \hat{\mathcal{G}}_{\mathbf{R}}\varrho(\mathbf{R}). \qquad (3.33)$$

In order to accomplish a basis-set independent notation these operators can be extended such that they project from any basis-set. For example, $\hat{\mathcal{R}}_{\mathbf{B}}$ would become the generalized real space projector $\hat{\mathcal{R}}$

$$\varrho(\mathbf{R}) = \hat{\mathcal{R}}\varrho(\mathbf{X}). \qquad (3.34)$$

Our modified Dirac notation bridges the gap between the formal Dirac language as used by physicists and a strongly basis-set dependent implementation as required in computer programs. The implementation simply provides functions like "toX()" in all vectors:

```
SxVector<Double> vEffR = vEffG.toR ();
```

Combining Dirac elements So far only binary Dirac operators have been discussed. An efficient implementation, however, requires sometimes to evaluate several Dirac operations simulataneously. Consider, for example, the evaluation of the kinetic energy in a plane-wave basis

$$E_{\text{kin}} = \langle \Psi_{\mathbf{G+k}} | \hat{L}_{\mathbf{G+k}} | \Psi_{\mathbf{G+k}} \rangle. \tag{3.35}$$

If only binary operations would be supported the evaluation would be inefficient, e.g., the following operations would be involved

$$t_{\mathbf{k}}(\mathbf{G}) = \nabla^2 c_{i\sigma\mathbf{k}}(\mathbf{G}) e^{-i(\mathbf{G+k})\cdot\mathbf{r}}$$

$$E_{\text{kin}} = \sum_{i\sigma\mathbf{k}} \sum_{\mathbf{G}} c^*_{i\sigma\mathbf{k}}(\mathbf{G}) e^{+i(\mathbf{G+k})\cdot\mathbf{r}} t_{\mathbf{k}}(\mathbf{G})$$

instead of the computationally less demanding direct evaluation according to

$$E_{\text{kin}} = \sum_{i\sigma\mathbf{k}} \sum_{\mathbf{G}} |c_{i\sigma\mathbf{k}}(\mathbf{G})|^2 |\mathbf{G+k}|^2. \tag{3.36}$$

In S/PHI/nX such combinations are accomplished by introducing container types which only store information about vectors or quantum numbers. For the above example an *empty* class SxLaplacian has been defined

```
class SxLaplacian { };
```

as well as a class describing the combination $\hat{L}|\Psi\rangle$

```
class SxLaplacianPsi {
   SxLaplacianPsi (Psi psiIn) : psi(psiIn) { }
};
```

This container class only stores a reference (p. 63) to the wave function, but does not compute anything. Due to reference counting this operation is fast. The remaining operation which applies $\langle\Psi|$ from left has been implemented similar to the operations introduced in Tab. 3.8, i.e., "(psi | SxLaplacianPsi)" will be mapped to the efficient function which evaluates Eq. (3.36). By applying this technique the kinetic energy reads in S/PHI/nX

```
SxLaplacian L;   // create empty laplacian
for (i=...)
   for (iSpin=...)
      for (ik=...)
         eKin += (waves(i,iSpin,ik) | L | waves(i,iSpin,ik);
```

This technique can be applied to evaluate any combination of Dirac expressions efficiently.

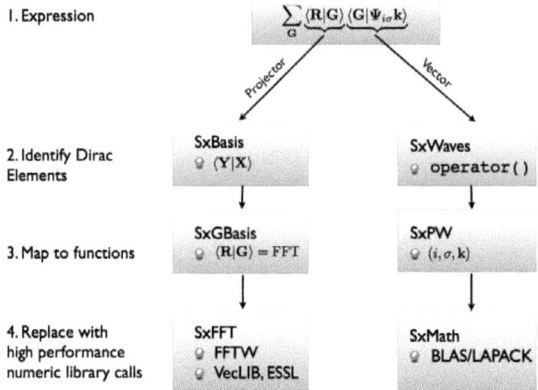

Figure 3.4: Demonstration of the compilation process by means of the implementation of $\langle \mathbf{G}|$ and $\langle \mathbf{R}|$ in order to support projectors like $\langle \mathbf{R}|\mathbf{G}\rangle$. When compiling $\langle \mathbf{R}|\mathbf{G}\rangle$ the virtual projector function of the basis-anchor $\langle \mathbf{Y}|\mathbf{X}\rangle$ redirects the compiler to the actual implementations (FFT functional call). Similarly, the term $\langle \mathbf{G}|\Psi_{i\sigma \mathbf{k}}\rangle$ is replaced by a vector handled by the SxMath class. Due to object-orientation the expressions $\sum_{\mathbf{G}} \langle \mathbf{R}|\mathbf{G}\rangle\langle \mathbf{G}|\Psi\rangle$ can be replaced by the proper FFT function call at compile-time.

The S/PHI/nX Dirac notation

In the previous paragraphs techniques to define projector elements $\langle \mathbf{X}|\mathbf{Y}\rangle$ between basis-sets and an interface to represent wave function coefficients $\langle \mathbf{B}|\Psi\rangle$ have been introduced. The missing element is to apply operators or projectors to wave functions. This has been accomplished by combining all previously discussed techniques, S/PHI/nX type mappers (p. 61), vector/matrix reference counting (p. 63), templates (p. 62), automatic BLAS/LAPACK function call mappings (p. 60), abstract basis-set/wave function classes (p. 77/78), braket templates (p. 80), virtual template projector functions (p. 81), and the automatic error detection (see Sec. 3.1.4) in a single function:

```
template<class Bra, class Ket>
operator* (SxProjector<Bra,Ket> proj, SxDiracVec<Ket::BasisType> vec)
{
   SX_CHECK (proj.ket);
   SX_CHECK (proj.ket == vec.basis));   // ''|A><B|'' is not allowed!
   return proj.ket->projectTo (proj.bra, vec, Ket::BasisType(0), Bra::BasisType(0));
}
```

This operator can be considered as the "glue" of the S/PHI/nX Dirac library. Note that this function does not generate machine code! Instead it is used by the compiler to create a highly efficient function call. The left argument ("`proj`") is replaced by a function while the right argument ("`vec`") is replaced by a suitable wave function coefficient vector. The operator can only create functions that are presented in Tab. 3.8 and apply them to available wave functions.

(a) Basis-set layer		(b) Wave function layer		(c) Dirac vector layer
Property	Example	Property	Example	
Mesh	$\mathbf{B}_i$	Basis-set ptr.	psi.getBasis()	-Algebraic vector
Integration	tr$_\mathbf{B}$	Access states	psi(i,s,k)	-Operators: +, -, *, /, ^, ...
Metric	$\langle a|b\rangle$			-Trigonometric functions
Projectors	$\langle \mathbf{B}|\mathbf{B}\rangle$,			-Pointer to basis-set
	$\langle \mathbf{X}|\mathbf{B}\rangle$			-Quantum numbers
	$\hat{\mathcal{R}}, \hat{\mathcal{G}}, \hat{\mathcal{B}}, ...$			-Memory management
Operators	$\hat{L}_\mathbf{B}, \hat{O}_\mathbf{B}, \nabla, ...$			-BLAS/LAPACK

Table 3.9: Organization of basis-set layer, wave function layer, and Dirac vector layer to construct the backbone of the Dirac notation at source code level. (a) The basis-layer defines general properties of a single basis-set, such as the mesh sampling and integration, scalar products with metrics or projectors onto other basis-sets. When adding new basis-sets only these features have to be overloaded in the new class. (b) Wave function classes are nothing but containers of Dirac vectors. The most important function is how to extract a single state. This function returns a Dirac vector. (c) Dirac vectors are algebraic vectors with a similar functionality as offered by high level toolkits such as Mathematica. Besides the numerical capabilities of these vectors information about the original basis-set and its quantum numbers can be retrieved. Our Dirac vectors are equipped with an automatic memory management.

The compilation process of an expression like

$$\sum_\mathbf{G} \langle \mathbf{R}|\mathbf{G}\rangle\langle \mathbf{G}|\Psi\rangle \qquad (3.37)$$

is illustrated in Fig. 3.4. Now it becomes clear, why an identity operation $\mathbf{1_B}$ (Eq. (3.31)) is useful. If the type of the vector psi is not known, the identity allows an explicit projection "SUM (G, (R|G) * (G | psi)". Assuming psi is a wave function in **G** space already. In this case the term "(G | psi)" will be replaced with the identity call. Otherwise a projection to **G** is performed before the FFT function is executed. Therefore, the two expressions

    psiR = SUM (G, (R|G) * (G|psiG));

and

    psiR = (R|psiG);

will result in the identical machine code (with the same performance). The same technique can be used to apply all other elements of the Dirac notation such as projectors, operators, the trace, and scalar products with metrics (see Tab. 3.9).

The context of a quantum mechanical expression is now detected by the compiler. Terms that are not defined cannot be specialized and result in a compiler error message. This is crucial as unphysical quantum mechanical expressions are automatically detected and will not compile in our method, even if the formal syntax is correct. Therefore, the S/PHI/nX library is not only a class library for quantum mechanical expressions. *S/PHI/nX may be considered as a quantum mechanical meta language!*

Equation	S/PHI/nX source code
$T = -\frac{1}{2}\nabla^2 \|\tilde{\Psi}_{i\sigma\mathbf{k}}\rangle = \hat{L}\|\tilde{\Psi}_{i\sigma\mathbf{k}}\rangle$	`T = L	waves(i,s,ik);`
$\tilde{\Psi}_{i\sigma\mathbf{k}}(\mathbf{R}) = \langle\mathbf{R}\|\tilde{\Psi}_{i\sigma\mathbf{k}}\rangle$	`psiR = (R	waves(i,s,ik));`
$\tilde{\Psi}_{i\sigma\mathbf{k}}(\mathbf{R}) = \sum_{\mathbf{G}+\mathbf{k}}\langle\mathbf{R}\|\mathbf{G}+\mathbf{k}\rangle\langle\mathbf{G}+\mathbf{k}\|\tilde{\Psi}_{i\sigma\mathbf{k}}\rangle$	`psiR = SUM(Gk,(R|Gk) * (Gk|waves(i,s,ik)));`
$\tilde{\Psi}(\mathbf{R}) = \sum_{\mathbf{G}+\mathbf{k},r}\langle\mathbf{R}\|\mathbf{G}+\mathbf{k}\rangle\langle\mathbf{G}+\mathbf{k}\|r\rangle\langle r\|\mu\rangle$	`psiR = SUM(Gk,SUM(r,(R|Gk)*(Gk|r)*(r|mu)));`
$\varepsilon_{i\sigma\mathbf{k}} = \langle\tilde{\Psi}_{i\sigma\mathbf{k}}\|\hat{H}\|\tilde{\Psi}_{i\sigma\mathbf{k}}\rangle$	`eps = (waves(i,s,k)	H	waves(i,s,k));`
$e_{\text{loc}}^{\text{ps}} = \langle V_{\text{loc}}^{\text{ps}}\rangle = \text{tr}(\rho V_{\text{loc}}^{\text{ps}})$	`eLocPS = tr(rho*vLocPS);`
$\varrho(\mathbf{R}) = \hat{\mathcal{R}}\varrho(\mathbf{G})$	`rhoR = rhoG.toR ();`
$\varrho(\mathbf{R}) = \hat{\mathcal{G}}\varrho(\mathbf{G})$	`rhoG = rhoR.toG ();`
$\varrho^{\text{ps}}(\mathbf{R}) = \hat{\mathcal{R}}\hat{\mathcal{G}}\varrho^{\text{ps}}(r)$	`rhoR = rhoPsRad.toG().toR();`

Table 3.10: Demonstration of C++ source code using the S/PHI/nX Dirac notation. The source code is strongly reminiscent to the original Dirac's notation in quantum mechanics. Note, that there is no drop in computational efficiency compared to a corresponding FORTRAN program.

With the high abstraction level of the source code the importance of the previously discussed automatic error detection (Sec. 3.1.4) becomes more pronounced. For example, consider an expression operating on two wave functions which have mismatching coefficient vector sizes (e.g. wave functions belonging to different basis-sets):

```
SxRBasis R (...);
SxGBasis G1 (eCut);
SxGBasis G2 (2 * eCut);
SxPW psi1 (G1);            // <G1|psi>
psiR = (R | psi1);         // < R|psi>
psi  = psi1 + ( G2 | psiR );  // (*) error detected here
```

The two vectors which are added in the last line have different number of vector elements which triggers the error detection mechanism in `SxVector<T>::operator+`. The generated error message, however, does *not* refer to `SxVector<T>::operator+` since this information would not be helpful. Instead S/PHI/nX's error detection identifies the calling source line "(*)" and can, thus, identify immediately the semantic error. In S/PHI/nX errors are detected in low-level routines but the messages refer typically to high level calling functions. This mechanism allows S/PHI/nX developers to locate and remove most of the typical semantic errors easily during the development process.

In Tab. 3.10 typical equations applied in Hamiltonians are shown. With our new meta language the previously sketched Hamiltonian can be now formulated almost like in textbooks. The Hamiltonian in the quasi-Dirac notation becomes in S/PHI/nX as simple as follows:

```
SxGBasis G;
SxGkBasis Gk;
SxRBasis R;
SxRhoG rhoG;
SxRhoR rhoR = rhoG.toR();
SxArray<SxDiracVec<Complex16> > T = G.getT ();
```

The Hartree potential/energy contributions can now be very intuitively expressed

```
for (is = ...)
   rhoGaussG += T(is) * SUM(r, (G|r)*(r|gaussianFunc));  // Eq. (1.62)
vHartree = FOUR_PI/G.gVec(SxIdx(1:ng)) * rhoGaussG;      // Eq. (1.65)
eHartree = 0.5 * tr (rho * vHartree);                    // Eq. (1.66)
```

as well as the local pseudo potential/energy contributions:

```
for (is = ...)
   vLocG += T(is) * SUM(r,(G|r)*(r|locPsFunc);   // Eq. (1.81)
eLocPs = tr (rhoR * (R|vLocG));                  // Eq. (1.82)
```

Also the previous example of the gradient can be encoded in a single transparent source line

```
...
dPsiG += (Gk(ik) | ( vEffR(iSpin) * (R|psiG) ); // Eqs. (2.5), (2.7)
...
```

and last but not least the expression to obtain the single particle energy (Eq. (2.10)):

```
double eps = ( psiG | H | psiG );
```

The application of our new techniques in quantum mechanical program packages combine various advantages:

- The source code becomes very short and compact. Algebraic and quantum mechanical expressions require only 1 or 2 source lines. The entire DFT Hamiltonian in S/PHI/nX could be implemented in less than 550 source lines.

- The functional approach makes the source code very transparent. Instead of reading many parts or files the algorithm can be understood in short time.

- The generated machine code is very efficient due to consequently mapping to highly efficient BLAS / LAPACK calls whenever possible.

- Numerics and computer related issues (like memory management) are strictly separated from physics.

- Future code extension are simple due to the modular concept.

- The implementation process is significantly faster than in conventional approaches. A sophisticated debug environment ensures that unphysical algebra equations or Dirac expressions do not compile. Physical and mathematical identities are validated at runtime to detect errors automatically. The error detection is able to identify the calling procedure that causes the problem.

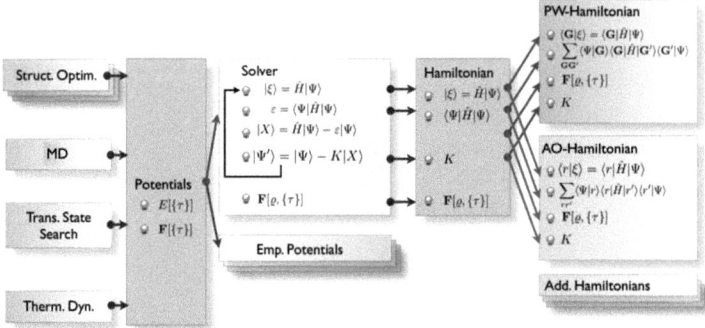

Figure 3.5: Fast development and execution speed requires often a mixture of generalization (grey shaded boxes) and conventionally implemented algorithms (colored boxes). The S/PHI/nX concept supports both. While the generalized form is usually very short, flexible, and simple to maintain, the conventional approach is best suited for method development and testing. From left to right: Structure related methods obtain their input from an abstract potential class. That can be either a DFT Born-Oppenheimer solver or any emperical potential. The Born-Oppenheimer solver minimizes the energy of an abstract basis-set independent Hamiltonian or, alternatively, of manually implemented basis-set dependent Hamiltonians.

3.2.4 Example

The benefit of a Dirac-like implementation can be best presented by means of a realistic code. In Tab. 3.10 the 1:1 relationship between expressions given in Dirac notation and the transcription into source code using our library is presented. The chosen equations are typical expressions when developing DFT codes.

As an example consider a code that projects an atomic orbital $\mu_{i_s i_a n k l m}$ to real space and computes the partial density afterwards in order to visualize the result $\varrho_i(\mathbf{R}) = |\langle \mathbf{R}|\mu_{i_s i_a n l m}\rangle|^2$. In the given example a direct projection $\langle \mathbf{R}|r\rangle$ is not defined. However, as both $\langle \mathbf{R}|\mathbf{G}\rangle$ and $\langle \mathbf{G}|r\rangle$ are defined in S/PHI/nX the partial density can be evaluated according to

$$\varrho(\mathbf{R}) = |\sum_{\mathbf{G},r} \langle \mathbf{R}|\mathbf{G}\rangle \langle \mathbf{G}|r\rangle \langle r|\mu\rangle|^2. \tag{3.38}$$

In Tab. 3.11 the fully functional and compilable source code is presented. As can be seen from the source code, this equation can be written as a single line in the code. As the library is based on the previously mentioned BLAS/LAPACK interface (SxMath) the executable's performance is very high.

3.3 Class Hierarchy

In the previous section abstract coding techniques have been derived which allow a high-performance implementation of quantum mechanical expressions. A critical design criteria of the S/PHI/nX library was that it provides a flexible basis to work with various quantum mechanical Hamiltonians (see Sec. 1.5), Hamiltonian derivatives (e.g. Density functional perturbation theory [106, 114]), new exchange-correlation functionals

Algorithm	Source code

	`#include <SPHInX.h>`
	`int main ()`
	`{`
	`  // — read input file`
	`  SxParser parser;`
	`  SxParser::Table input = parser.read ("input.sx");`

read $\{\tau_{i_s i_a}\}$ read $\{\phi^{\text{ps}}_{i_s i_a nl}(r)\}$	`  SxAtomicStructure str (input);` `  SxPseudoPot psPot (input);`

setup N_1, N_2, N_3
read E_{cut}
	`  // — setup FFT mesh resolution and energy cut-off`
	`  SxVector3<Int> mesh (SxGBasis::getMesh(input));`
	`  double gCut (SxGBasis::getEcut(input));`

$\langle \mathbf{R}\|$
$\langle \mathbf{G}\|$
$\langle \mathbf{G}+\mathbf{k}\|$
$\langle r\|$
$\langle r\|\mu\rangle = \phi^{\text{ps}}_{i_s i_a nl}(r) y_{lm}$
	`  // — build Dirac basis-sets and atomic orbitals`
	`  SxRBasis R (mesh, str.cell);`
	`  SxGBasis G (mesh, str, gCut);`
	`  SxGkBasis Gk (G, input);`
	`  SxRadBasis r (psPot, str.cell);`
	`  SxAtomicOrbitals mu(psPot, r);`

$\left\| \sum_{\mathbf{G}} \langle \mathbf{R}\|\mathbf{G}\rangle \langle \mathbf{G}\|\mu_{i_s,i_a,n,l,m}\rangle \right\|^2$	`  // — compute partial density`
	`  int is=0, ia=0, n=0, l=1, m=0;`
	`  R.writeMesh3d ("s.sxb",`
	`    SUM (G, (R|G)*(G|mu(is,ia,n,l,m))).absSqr()`
	`  );`
	`  return 0;`
	`}`

Table 3.11: Demonstration of implementing an algorithm using the S/PHI/nX Dirac notation capabilities. The presented code is not a pseudo code, but a complete and compilable source code!

(e.g. Exact exchange formalism [115]). To achieve this aim, also the electronic minimizers (such as steepest-descent see Sec. 2.1.2, or conjugate gradient schemes, see Sec. 2.1.4) have to be flexible enough to allow an straight-forward modification in future.

3.3.1 Electronic minimization

In Sec. 2.1 a brief overview about *state-of-the-art* schemes to diagonalize the Hamiltonian iteratively was presented. The number of iterations n^{it} which are necessary to reach the Born-Oppenheimer surface within the required accuracy and, in particular, the number of evaluations of the gradient $\frac{\delta \hat{H}}{\langle \delta \Psi |}$, i.e., Eqs. (2.5), (2.6), (2.7)), determines strongly the performance of a DFT program package. In order to increase the computational performance the identification and/or development of algorithms that minimize the number of iterations n^{it} is critical. Furthermore, the minimization schemes differ by the number of temporary wave function objects which increases the memory demand (see, e.g., Sec. 2.1.4). Therefore, the source code dedicated to converge down to the Born-Oppenheimer surface is typically modified very frequently. In order to address the issues of performance and memory consumption of the minimization schemes in S/PHI/nX various methods to simplify testing and implementation of advanced schemes have been developed. That are methods to (i) strictly separate Hamiltonian / potential sources from that of the the multi-dimensional minimization algorithms using automatic pointers, (ii) modular support of preconditioners as used in conjugate-gradient schemes, (iii) a simultaneous support of vector and matrix representations, and (iv) a test environment for iterative schemes. These methods will be discussed in the following.

Automatic pointers

As discussed above in S/PHI/nX a functional programming approach has been chosen instead of procedural programming (see Sec. 3.1.4). On one hand functional programming increases the transparency of the code drastically, on the other hand various problems ranging from performance drops to data type handling had to be solved. In the previous section (pp. 62) it was shown how to overcome the drawbacks of functional programming. We will describe in this section how an efficient functional programming can be accomplished for obtaining the Born-Oppenheimer surfaces as well.

A key element for the iterative minimization schemes introduced in Sec. 2.1.1 is the gradient Eq. (2.3) . By overloading the "|" operator, the evaluation of $\hat{H}|\Psi\rangle$ reads in S/PHI/nX as follows:

```
dPsi = H | psi;
```

The choice of which Hamiltonian should be used is not known at compile time. When performing a calculation with S/PHI/nX the Hamiltonian should be chosen from an input file. Hence, the data types of H and psi are entirely unknown for the compiler. As introduced in Sec. 3.2.3 C++ provides placeholder functions (virtual functions) which can be applied here. The usage of virtual functions requires, however, that H and psi cannot be static variables any longer. Instead C++ expects them to be pointers to the corresponding class. For example, a generalized Hamiltonian class would provide an interface

```
class SxHamiltonian
{
    ...
```

```
        virtual Psi operator| (Psi);
        virtual double getEnergy (Rho);
};
```

The corresponding Hamiltonian class can be derived from `SxHamiltonian` and provides the virtual functions `operator|` and `getEnergy`. The decision about the choice of the Hamiltonian can then be done at run-time:

```
        // initialize 'hType' from input file
        ...
        SxHamiltonian *H = NULL;    // create pointer to abstract Hamiltonian
        if (hType == PW)    H = new SxPWHamiltonian;
        if (hType == Test)  H = new SxTestHamiltonian;
        ...
        Psi dPsi = H | psi;
        ...
        delete H;   // free memory
```

This approach is very error-prone. Here, C pointers have to be applied in the physics part of S/PHI/nX. While C pointers provide a huge degree of flexibility (see Sec. 3.1.3) they are also very dangerous. Even minor mistakes in their handling can create catastrophic problems, ranging from memory leaks to an unpredictable run-time behavior. The difficult and error-prone usage of C pointers is one of the main hurdles for beginners in C/C++. Hence, a main goal in the development of the S/PHI/nX library was to avoid a direct handling with C pointers at the physics level. A safer way of providing the required flexibility had to be found.

Various modern computer languages (such as Java or ObjectiveC) tackle the problem of pointer misusage with *garbage collecting*. In this approach an external process constantly searches for unused memory resources and releases them. This approach is not suitable for high-performance computing (HPC): The garbage collecting task consumes typically 5-10% of the CPU load which was not acceptable to us. Furthermore, resources can be released with a considerable delay. When releasing huge objects (such as the memory for large matrices) the memory might be required instantly for the subsequent operation. The approach of garbage collecting was, therefore, no alternative for our approach.

In S/PHI/nX we, therefore, combine the reference counting technique which has been applied to reduce the memory accesses in algebraic expressions (p. 63), C++ template techniques (p. 62), and automatic error detection (see Sec. 65) to create automatic pointers. An automatic S/PHI/nX pointer is a template class for any data type <T> which wraps all memory accesses. The reference counting ensures that the memory will be automatically deleted when no object refers to the pointer anymore. The above pseudo code becomes

```
{
        ...
        // --- read from input file
        SxPtr<SxHamiltonian> H; // create automatic pointer to abstract Hamiltonian
        if (hType == PW)    H = SxPtr<SxPWHamiltonian>::create ();
        if (hType == Test)  H = SxPtr<SxTestHamiltonian>::create ();
        ...
        Psi dPsi = H | psi;
```

```
...
}    // H will be released here automatically
```

Every memory access of `SxPtr<T>` is monitored by the S/PHI/nX error detection mechanism and every memory violation is detected immediately. The S/PHI/nX autopointer provides the same flexibility as C pointers. No CPU-time consuming garbage collecting process is necessary. The memory is automatically released without any delay.

The automatic S/PHI/nX pointers provide a simple way of generalizing Hamiltonians as well as potentials and, consequently, to decouple the source code from the specific electronic minimization schema. The above techniques to generate abstract and modular interfaces have been introduced in S/PHI/nX for Hamiltonians and electronic minimization schemes to obtain the Born-Oppenheimer surfaces. Beside the DFT potential also (semi-)empirical potentials are available in S/PHI/nX. All algorithms related to atomic structures (structure optimization, molecular dynamics, transition-state search) access an abstract potential class. The required techniques to create such general interfaces have been discussed in the previous section (p. 77/78).

Vector/matrix representation

In Sec. 2.1.4 the highly efficient all-band conjugate gradient and state-by-state conjugate gradient minimization schemes have been introduced. The requirements for an implementation of both are very different. The all-band conjugate gradient scheme requires a matrix formulation while the state-by-state conjugate gradient method uses a vector form. This applies to wave functions $\Psi_{i\sigma \mathbf{k}}/\mathbf{C}_{i\mathbf{G}}(\sigma \mathbf{k})$, the gradients $|\xi_{i\sigma \mathbf{k}}\rangle/\Xi_{i\mathbf{G}}(\sigma \mathbf{k})$, as well as the preconditioners $K(|\Psi_{i\sigma \mathbf{k}}\rangle)/K(\mathbf{C}_{i\mathbf{G}}(\sigma \mathbf{k}))$. Following previous discussions from Sec. 3.2.2 both representations are already supported in our approach for the first two cases. Only a simultaneous application of both preconditioner forms has to be supported by the Hamiltonian class, i.e.,

```
PsiG    SxHamiltonian::preconditioner (PsiG);
PsiGI   SxHamiltonian::preconditioner (PsiGI);
```

The first function represents the preconditioner in vector form, the second function is the matrix counterpart. The preconditioner source code can be decoupled from that of the electronic minimization library using automatic pointers (Sec. 3.3.1) to the Hamiltonian, i.e., `SxPtr<SxHamiltonian>`.

The importance of a preconditioner is illustrated in Fig. 3.6 where the performance of the conjugate gradient scheme is compared with an optimized steepest descent[36] (Sec. 2.1.2). In both cases the tests have been conducted with and without preconditioners. The steepest descent with quadratic line minimization (optimized steepest descent) is shown as dotted line. The convergence rate is not logarithmic. Only *conjugating* the search directions can reduce the dimensions of the search vector space in every iteration and a truly logarithmic convergence rate can be accomplished. In both cases the application of a preconditioner can significantly improve the convergence behavior. As illustrated, a speed up factor of two can easily be accomplished. In the performed test an Arias preconditioner has been applied. It can be seen that the application of preconditioning is crucial to achieve optimal performances.

In the current implementation of S/PHI/nX, for the conjugate gradient schemes there are two preconditioners available, the Payne preconditioner [61, 116]

$$K = \frac{27 + 18x + 12x^2 + 8x^3}{27 + 18x + 12x^2 + 8x^3 + 16x^4} \qquad x = \frac{|\mathbf{G} + \mathbf{k}|^2}{\mathbf{g}_{\mathbf{G}i}} \qquad (3.39)$$

---
[36]steepest descent with line minimization

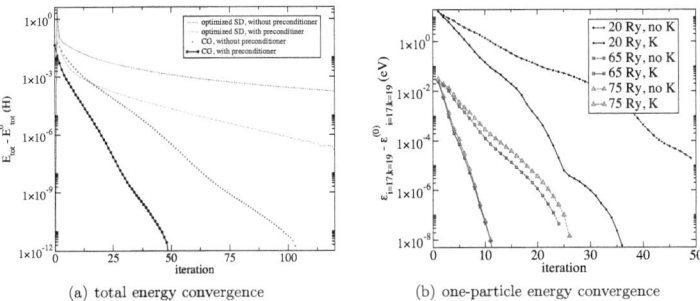

(a) total energy convergence (b) one-particle energy convergence

Figure 3.6: Influence of the contributions to a preconditioned conjugate gradient scheme in case of ZnO bulk to the (a) total energy convergence and (b) one-particle energy convergence of a randomly chosen state at different energy cut-offs and a **k**-point folding of $[\frac{1}{2}\frac{1}{2}\frac{1}{2}] \times \{4,4,4\}$. For each iteration the difference of the total energy E_{tot} or one-particle energy ε to the converged solution E_{tot}^0 or ε^0 are given.

as well as the Arias preconditioner [67]

$$K = \frac{\sum_{i=0}^{8} x^i}{\sum_{i=0}^{9} x^i} \qquad x = \frac{|\mathbf{G}+\mathbf{k}|^2}{E_{\text{kin}}/n} \qquad (3.40)$$

with n denoting the number of bands, $\mathbf{g}_{\mathbf{G}i}$ refers to the vector of **G** components of the gradient vector defined in Eq. (2.22). Both preconditioners are constructed by polynoms depending on the kinetic contribution. The denominator is a polynom one order higher than the nominator in order to remove the $|\mathbf{G}+\mathbf{k}|^2$ term. The Payne preconditioner depends on the kinetic energy per state whereas Arias is using an averaged kinetic contribution. In Fig. 3.7 the general shape of both preconditioners is compared. The Arias and Payne preconditioner yield comparable convergence rates.

In Fig. 3.8 the residue vector $|\Psi(\mathbf{G}) - \Psi_0(\mathbf{G})|^2$ obtained with and without application of a preconditioner is plotted in the spectral representation. If the residue vector vanishes for all **G** the solution has been found. As illustrated in Fig. 3.8 the high frequency contributions of the residue vector vanish when the preconditioner is applied. Therefore, the dimension of the search space can be significantly reduced which leads to a dramatic increase of the convergence rate.

In Sec. 2.1.4 the subspace rotation has been introduced in Eq. (2.34). For systems that can be described with only fully occupied states (such as insulators or semiconductors), the application of a subspace rotation [67, 27] seems not to be important at first. As can be seen from Fig. 3.9 the application of a subspace rotation can significantly improve the convergence speed and stability in case of semiconductors. Hence, in S/PHI/nX the subspace diagonalization is applied for all systems by default.

Orthonormalization In the previously discussed minimization schemes the orthonormalization constraint is fulfilled by a Gram-Schmidt orthonormalizer. Here each state is sequentially orthogonalized to all energetically lower lying states. This scheme can also be formulated using blocked matrix-operations. The matrix oriented orthonormalization method is known as Löwdin orthonormalization. It creates $\mathbf{C}_\perp$ which is the

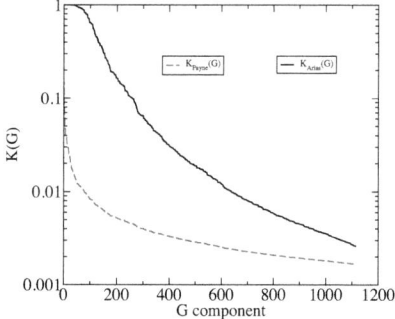

Figure 3.7: Shape of both Payne and Arias preconditioners during the electronic minimization cycle for a ZnO-bulk with 65 Ry cut-off. Both preconditioners damp the high frequency components of the search direction vector. In order to keep the norm of the search vector untouched the **G**=**0** component of the preconditioner **K**(**G**=**0**) is 1. The numerical fluctuations are due to the kinetic contribution of the search vector itself.

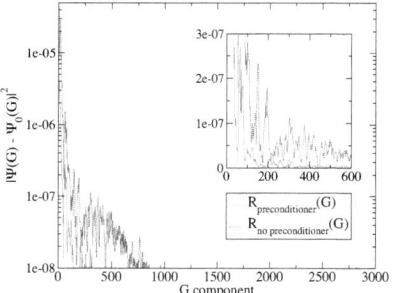

Figure 3.8: Effect of preconditioning in case of a ZnO-bulk with 65 Ry energy cut-off. In this figure the **G** frequency resolved spectrum of the residual wave function $|\Psi(\mathbf{G}) - \Psi_0(\mathbf{G})|^2$ is depicted. $\Psi(\mathbf{G})$ is the wave function of a randomly chosen state after performing an iteration of a conjugate gradient method. $\Psi_0(\mathbf{G})$ refers to the fully converged wave function belonging to the same state. The solid line depicts the result obtained with an Arias-like preconditioner whereas the dotted line shows the same situation without the application of a preconditioner. In both cases the error wave function has the strongest contribution in the region of small **G** vectors. Hence, the time consuming part of the minimization is due to the long-range contributions. In case of an high energy cut-off system (here 65 Ry) the preconditioner can decrease the error wave function significantly and hence the convergence rate becomes improved. The inset magnifies the low-frequency region on an linear scale.

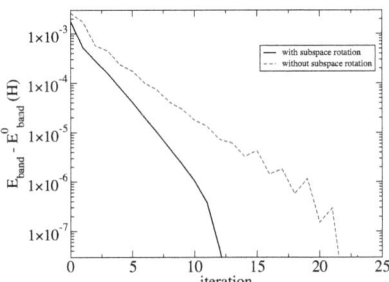

Figure 3.9: Using the state-by-state conjugate gradient scheme for a band structure calculation of a semiconducting system (ZnO-bulk). The diagram displays the convergence rate of the band energy $E_{\text{band}} = \sum_{i\sigma\mathbf{k}} \omega_{\mathbf{k}} f^{\text{occ}}_{i\sigma\mathbf{k}} \varepsilon_{i\sigma\mathbf{k}}$. Note that the total energy E_{tot} is not variational when the density is kept fixed and hence, E_{tot} cannot be used to analyze the convergence behavior. The energy axis is scaled according to the converged band energy value E^0_{band}. The picture also shows that the application of the subspace rotation is also useful for semiconducting systems. Without subspace diagonalization the numerical error is energy dependent. Energetically higher lying states show a larger numerical error and converge therefore slower. When the numerical accuracy has been reached H is not entirely diagonalized.

matrix containing the orthonormalized wave function coefficients by applying an uniform transformation **U**.

$$\mathbf{S} = \tilde{\mathbf{C}}^\dagger \tilde{\mathbf{C}} \qquad (3.41)$$

$$\mathbf{S}\mathbf{v} = s\mathbf{v} \qquad (3.42)$$

$$\mathbf{U} = \mathbf{v}^\dagger \frac{1}{\sqrt{s}} \mathbf{v} \qquad (3.43)$$

$$\mathbf{C}_\perp = \mathbf{U}\tilde{\mathbf{C}}. \qquad (3.44)$$

The implementation in S/PHI/nX is as simple as

```
I   = S.identity ();
eig = S.eigensystem ();                                    // Eq. (3.42)
U   = eig.vecs.adjoint() ^ (I/sqrt(eig.vals)) ^ eig.vecs;  // Eq. (3.43)
psi = U ^ psi;                                             // Eq. (3.44)
```

Please note, that SxMath internally computes the optimal data types (see Sec. 3.1.4) depending on the matrix shape and type of **S** and automatically applies blocking (see Sec. 3.1.3). While the source code remains very close to the actual equations, the executional performance is very high.

The usage of the Löwdin orthonormalization has two significant advantages over the Gram-Schmidt counterpart. It is very efficient as it is based only on blocked operations and can, therefore, exploit modern computer architectures most efficiently. Numerically it is also more stable than the Gram-Schmidt scheme because numerical errors are uniformly distributed to all states equally. In the Gram-Schmidt scheme the numerical error increases with higher lying states.

RMM-DIIS The final element in providing an efficient interface to implement efficient minimization techniques is the support of charge density mixing schemes (see p. 43). The S/PHI/nX libraries and the functional programming approach allow a straightforward implementation of the RMM-DIIS mixing scheme:

```
for (i=0; i < m; ++i) {
  for (j=i; j < m ; ++j)
    A(i,j) = (dR(i) | dR(j));      // Eq. (2.51)
  B(i) = (dR(i) | R(m));           // Eq. (2.53)
}
alpha = -(A.inverse() ^ B);        // Eq. (2.52)
M = g2 / (g2 + q0);                // Eq. (2.49)
K = rho + (M * rho.toG()).toR ();  // apply metric in <G| space
rhoOpt  = K*R(m);                  // Eq. (2.54)
for (i=0; i < m-1; ++i)
  rhoOpt += alpha * (dRhoIn(i) + K*dR(i));
```

In order to increase the source code transparency, in S/PHI/nX the preconditioner K is defined as an operator which can be applied on charge densities or residual vectors via the "*"-operator.

Linearized gradient test

The iterative nature of the electronic minimization schemes introduces a difficulty when developing/modifying the Hamiltonian or potential classes. During the development process a significant time is spent in testing the modified algorithms. In case of iterative minimization schemes this is, however, challenging. The result can only be verified after full convergence has reached. In case of inconsistencies the result will converge to an unphysical value and locating the source of the problem is not trivial. This problem also applies to the identification of the origin of numerical inaccuracies. In numerics various expressions can introduce losses in the accuracy, e.g., division of large values by small numbers [92]. After the SCF cycle the identification of the source line where such numeric instabilities have been introduced is difficult.

In order to address these two important issues, in S/PHI/nX a test environment to verify the consistency of potential contribution and its corresponding gradient has been developed. This linearized gradient test environment is demonstrated in Fig. 3.10. In this example the performance of a *quadratic* line minimization (see Sec. 2.1.4) is analyzed to test whether a second order fit using one trial energy $E_{trial}^{(n)}$ and the derivative Eq. (2.26) is sufficient. While the incorporation of higher orders might lead to a better sampling of the true energy function, more total energy values or derivatives need to be provided. Every energy value or derivative is computationally demanding[37]. From Fig. 3.10 one can see that even far off the minimum the prediction of λ_{min} is only 20% smaller than the true minimum. With every subsequent iteration the prediction is close to the true value. Hence, for our applications higher order fits would only waste CPU time.

Such tests are typical for the development process. In S/PHI/nX any part of the Hamiltonian can be tested *non-iteratively* and separately for single states using the linearized gradient test. Numerical inconsistencies between the analytic energy expression and the numerical linearized gradient can be determined easily. By applying the linearized gradient test for every contribution to the Hamiltonian numerical instabilities could be easily identified. The highest *numerical* accuracy of the total energy in S/PHI/nX is currently[38] $\Delta E_{tot} \approx 1e^{-12}$ H. This is equivalent to the estimated achievable numerical accuracy[39].

---
[37]It requires an update of the Hamiltonian $\hat{H}[\varrho]$ and an orthogonalization step!
[38]The measurements have been performed on AMD Opteron 246, 64-bit, using GNU C++ compiler.
[39]The achievable accuracy can be *roughly* estimated [92] depending on the involved operations: data type "double" = 16 digits. $\Delta(+,-) = -1$ digit, $\Delta(*,/,\mathrm{sqrt}) = -2$ digits, $\Delta(\sin,\cos,\exp) = -4$ digits.

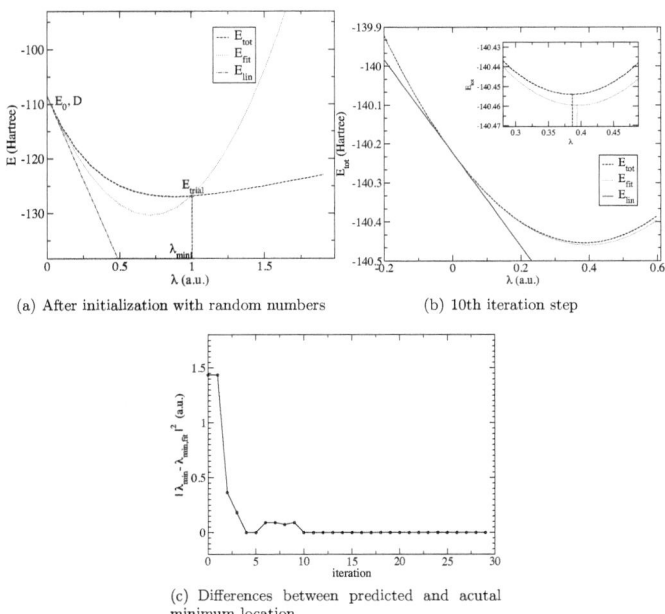

(a) After initialization with random numbers
(b) 10th iteration step
(c) Differences between predicted and acutal minimum location

Figure 3.10: Quality of the quadratic line minimization demonstrated by means of a ZnO bulk analyzed using the linearized gradient test with the total energy E_{tot}, the fitted energy E_{fit} (Eq. (2.27), and the linearized energy E_{lin} (Eq. (2.26)). (a) Far off the minimum the predicted energy value does not fit well the actual total energy value. However, applying the predicted minimum λ_{trial} improves the wave function significantly. Each iteration step leads to a new starting value that is closer to the quadratic regime. Therefore the quadratic line minimization predicts the minimum better with every step. (b) After performing a few iterations the fitted total energy $E_{\text{fit},0}$ is still not predicting the real total energy minimum (see Fig. (b) inset). However, the discrepancy between λ_{trial} and the true λ is very small. Higher order fits could not significantly improve the predicted λ and would, therefore, not justify additional (expensive) total energy calculations. The closer the wave functions are to the self-consistent solution the better the quadratic fit. Hence, the scheme stabilizes itself. (c) The difference between the real and the predicted location of the minimum vanishes rapidly with progressing conjugate gradient steps.

3.3.2 Representing atomic structures

In a plane-wave package the majority of the CPU time is spent in the computation of the Born-Oppenheimer surface (Sec. 2.1) and the Hellmann-Feynman forces (Sec. 1.7.1). The time which is required to move atoms along the force gradients to relax atomic structures or to perform molecular dynamics (Sec. 2.2.2) can be neglected compared to the minimization of the total energy. However, the S/PHI/nX project aims to be a basis-set independent program and many different potential types should be applicable within this framework, ranging from computationally expensive DFT potentials to fast empirical potentials such as Stillinger-Weber or Lennard-Jones potentials [1]. Empirical potentials can easily be applied to systems with thousands or tens of thousand of atoms [4]. The evaluation of the forces with these potentials is very fast. Therefore, when developing/implementing structure optimization algorithms into S/PHI/nX, performance is a most crucial issue. In order to cope with this performance problem while providing a similarly intuitive library like the S/PHI/nX DFT library we introduce in this section a concept of how atomic structures can be represented efficiently with respect to CPU time while providing an intuitive programming interface.

Modular description of atomic structures and forces

The equations used in structural related algorithms, such as structure optimization [99], transition state searches [1], phonon calculations (Sec. 2.3), or molecular dynamics are often based on Newton's mechanics. Most of these schemes involve rather simple algebraic equations. Yet actual implementations in modern programs tend to be large. Consider for example a simple damped Newton algorithm. The atom structure, given as a set of atomic positions $\{\tau\}$, can be relaxed according to [61]

$$\tau^{n+1}_{i_s i_a d} = (1 + \lambda_{i_s})\tau^{n}_{i_s i_a d} - \lambda_{i_s}\tau^{n-1}_{i_s i_a d} + \mu_{i_s} F_{i_s i_a d} \quad \text{with } d = (xyz). \quad (3.45)$$

Here $\tau_{i_s i_a}$ denotes the atomic position of the atom i_a belonging to the species i_s. λ and μ are species dependent convergence parameters. λ can be interpreted as a damping parameter and μ as a reduced mass. $F_{i_s i_a d}$ is the dth component of the force vector. The previous, current, and new atomic structures are identified by their iteration numbers $n-1$, n, and $n+1$, respectively. The implementation of such expressions requires extensive loops over the indices i_s, i_a, and d. Some representative example code code read as follows:

```
for (is=0; is < nSpecies; ++is) {
   for (ia=0; ia < nAtoms(iSpecies); ++ia) {
      for (d=0; d < 3; ++d) {
         tauNew(is,ia)(d) = (1. + lambda(is)) * tau(is,ia)(d)
                          + lambda(is) * tauPrev(is,ia)(
                          + mu(is) * F(is,ia)(d);
         tauPrev(is,ia)(d) = tauNew(is,ia);
      }
   }
}
```

Following the discussions in the previous section the resulting machine code is inefficient: numerical operations on the arrays tau, lamda, and F are inefficient because software pipelining/look-ahead mechanisms (p. 58) cannot be applied. Even more important, this example code does not benefit from matrix blocking algorithms (Sec. 3.1.3). Besides the efficiency, the development of an intuitive source code has been and is always a key item in the S/PHI/nX project. The required source lines for loops and index handling requires a detailed index-based implementation which can be cumbersome and error-prone[40]. In order to simplify Eq. (3.45) the terms are separated into *classes of variables*:

1. 3d coordinate vectors depending on (d, i_s, i_a), e.g., τ or F
2. parameters depending on the species i_s only, e.g., λ or μ
3. entities independent of species or atoms

For the first group the S/PHI/nX class `SxAtomicStructure` has been implemented which basically represents the atomic coordinates in terms of n_{species} coordinate matrices with dimensions $3 \times n_{\text{atoms}}$. These coordinate matrices are represented with `SxMatrix<T>` objects (see Sec. 3.2) which allows for an efficient BLAS3 mapping. The algebraic operators ('+', '-', '*', '/' etc.) must be defined such that they loop automatically over the proper indices, for example:

```
operator+ (SxAtomicStructure a, SxAtomicStructure b)
{
    // --- verify code consistency (see Sec. 3.1.4)
    SX_CHECK (a.getNSpecies() == b.getNSpecies());
    SX_CHECK (a.getNAtoms()   == b.getNAtoms());

    SxAtomicStructure res(...);
    for (is=0; is < a.getNSpecies(); ++is)
       res(is) = a(is) + b(is);    // mapped to BLAS3 matrix operation!
    return res;                    // using ref. counting (p. 63)!
}
```

Note that the expressions `res(is)`, `a(is)`, and `b(is)` are efficient matrix operations.

Loops of the second type of the above list (species dependent entities) can be implemented similarly:

```
operator* (SxVector<T> a, SxAtomicStructure b)
{
    // --- verify code consistency (see Sec. 3.1.4)
    SX_CHECK (a.getSize == b.getNSpecies());

    SxAtomicStructure res (...);
    for (is=0; is < b.getNSpecies; ++is)
       res(is) = a(is) * b(is);    // mapped to BLAS2 operation
    return res;                    // using ref. counting
}
```

---

[40]In this simple example 3 source lines are required for index loop handling for a single equation. For more complex algorithms considering case differentiations (discussed below) the ratio between source lines for index handling and actual expressions can become even worse.

With such an approach the extensive index handling can be completely avoided and the corresponding source code for implementing Eq. (3.45) becomes as simple as

```
SxAtomicStructure tauNew, tau, tauPrev, F;
SxSpeciesData lambda, mu;
while ( (tauNew-tau).absSqr().maxVal() > 1e-8 )  {    // convergence?
    F       = hamSolver.getForces (tau);             // independent of actual H
    tau     = tauNew;
    tauNew  = (1+lambda)*tau - lambda*tauPrev + mu*F; // Eq. (3.45)
    tauPrev = tauNew;
}
```

As for the electronic part (previous section), the numerical operations are mapped to the proper BLAS calls to guarantee peak performance.

Coordinate representation

Depending on the structure optimization schemes coordinates of atoms or forces must be treated differently. A very intuitive way of describing a coordinate is in the Cartesian system, e.g.,

$$\tau_{\text{cart}} = \begin{pmatrix} x \\ y \\ z \end{pmatrix}.$$

In other scheme, such as the quasi Newton scheme [99] which is described in the following section, a degree of freedom (DoF) representation of all coordinates at once is necessary, e.g.,

$$\tau_{\text{DoF}} = \begin{pmatrix} x_{i_a=1} \\ y_{i_a=1} \\ z_{i_a=1} \\ x_{i_a=2} \\ \vdots \\ z_{i_a=n_a} \end{pmatrix}.$$

Some algorithms even need to change between both representation repeatedly[41]. Due to efficiency considerations copying of the vector elements must be avoided. The previously sketched solution combines both representations. The internal storage using <T> allows a DoF representation while in S/PHI/nX functions applying reference counting (p. 63) extract Cartesian coordinates without copying elements.

Transformations

In structure optimization calculations it is often very useful to apply constraints. For example, when performing a relaxation of the surface layers only, the surface atoms should be relaxed whereas atoms belonging

---

[41] For example, structure relaxation of spatially constrained atoms (coordinate representation) with the BFGS algorithm [99] (DoF representation).

to the bulk region must be kept fixed. Furthermore, it should be possible in S/PHI/nX to constrain high-frequency atomic movements along specified directions. When computing molecular systems center of mass filters can be applied or rotations due to numerical noise have to be projected out. Typically those operations are intermixed with the structural algorithms such as

```
for (is=0; is < nSpecies; ++is) {
   for (ia=0; ia < nAtoms(is); ++ia) {
      if (moveAtom) {               // keep atoms fixed (''sticky filter'')
         if (applyCenterOfMass) {
            ...
         }
         if (applyHighFreq) {
            ...
         }
         if ...
      }
   }
}
```

The same filter operations have to be applied to all structural algorithms which lead to a large redundancy of source code fragments. In order to decouple such constraints from the actual multidimensional minimization algorithms, "S/PHI/nX transformation pipelines" have been introduced. In this approach one can define all transformations in terms of a new operator acting on $\tau_{i_s i_a}$ and/or $\mathbf{F}_{i_s i_a}$. Letting $\hat{T}$ be a general transformation, Eq. (3.45) would become

$$\tau^{n+1} = (1+\lambda)\tau^n - \lambda\tau^{n-1} + \mu\hat{T}\mathbf{F}. \qquad (3.46)$$

The actual form of $\hat{T}$ can be defined elsewhere, for example as user input from an input file. Of course, transformations have to be capable of combing like

$$\hat{T} = \hat{T}_1|\hat{T}_2|\ldots|\hat{T}_n. \qquad (3.47)$$

Every transformation can be modularly defined in a separate S/PHI/nX class:

```
class SxStickyFilter : public SxTransform
{
   ...
   operator* (SxAtomicStructure a) {
      SxAtomicStructure res = ...;
      return res;
   }
};
class SxCenterOfMass : public SxTransform
{
   operator* (SxAtomicStructure f) {
      return f - f.sum () / f.getNAtoms();
```

```
         }
     };
```

The base class SxTransform provides the pipeline mechanism which allows to decouple the source codes of transformations entirely from those of the minimization schemes:

```
// --- create dynamic pipeline of transformations
SxTransform T;
if (applyStickyFilter)  T = T | SxStickyFilter (...);
if (applyCenterOfMass)  T = T | SxCenterOfMass (...);

// --- decoupled structural minimization scheme
while ( (tauNew-tau).absSqr().maxVal() > 1e-8)  {    // convergence?
   F      = T | hamSolver.getForces (tau);          // apply transformation pipeline
   tau    = tauNew;
   tauNew = (1+lambda)*tau - lambda*tauPrev + mu*F; // Eq. (3.46)
   tauPrev = tauNew;
}
```

Quasi Newton

The quasi Newton scheme has been presented in Sec. 2.2.1. Using the previously discussed techniques for representing atomic structures in S/PHI/nX and the usage of transformation pipelines this algorithm can be implemented remarkably simple while the machine code is very efficient due to reference counting (p. 63, blocking (p. 68). This efficiency is important for large-scale calculations using empirical potentials.

```
tau -= B.inverse() ^ g.coordRef ();        // (A): convert DoF to cart. repr.
g    = T | getForces(x);                   // (B): apply transformation pipelines
s    = tau - tauOld;                       // Eq. (2.59)
y    = g.coordRef () - gOld.coordRef ();   // Eq. (2.60)

// --- (C): BLAS mapping
B -= B^s^s.transpose()^B.transpose()) / (s.transpose()^B^s)    // Eq. (2.61)
   - (y^y.transpose())               / (y.transpose()^s);
```

In this example all techniques described in this section have been applied. In line "(A)" the degree of freedom representation is converted without performing copy operations to the Cartesian representation. The forces are filtered/transformed in line "(B)". The DoF representation is mapped to BLAS calls in line "(C)".

3.3.3 Add-ons

So far we discussed the modular library aspects of the S/PHI/nX project. By including the above described classes simulation programs, analysis tools, or other preparation tools can be easily developed by including the S/PHI/nX class libraries. The executables built with S/PHI/nX are called S/PHI/nX add-ons. They provide a standardized command line interface. The output of one S/PHI/nX add-on can serve as input for another. This way S/PHI/nX add-ons are scriptable. Currently S/PHI/nX comes with a set of about 50 add-ons such as tools to setup complex atomic structures, analysis tools to compute (partial) densities of state (DOS), tools to operate on wave functions and/or charge densities/potentials as well as file i/o converters to connect the S/PHI/nX project to 3rd party tools.

The usage of the highly abstract S/PHI/nX library allows the add-ons to be remarkably short with respect to the number of source lines. Typically, even complex analysis operations could be programmed with 50-150 source code lines.

Since the IT market progresses rapidly with respect to the introduction of new hardware and operating systems, S/PHI/nX has been developed as cross-platform project. To simplify future ports to new architectures, S/PHI/nX has been and will be developed while strictly obeying the ANSI C++ standard which allows to use a wide range of C++ compilers. In order to communicate with the underlying operating system only a subset of POSIX functions has been applied. As a result S/PHI/nX is available on all major platforms, such as Linux, MacOS X, FreeBSD, AIX, HP-UX, Windows XP/Vista/7rc1.

3.4 Comparison with VASP

In the previous sections we introduced the S/PHI/nX approach in which quantum mechanical algorithms can be implemented in a formulation which is strongly reminiscent to the Dirac notation. This provides a number of advantages: new algorithms can be easily developed and tested and code maintenance becomes simple and the program package can be kept small. The ultimate question, however, is whether such advantages introduce performance penalties which, of course, would render the entire approach useless.

In order to estimate the actual performance of the S/PHI/nX approach it is necessary to compare the run-time performance of realistic simulations with other standard-codes. In the scope of this work thermodynamic properties of III-V semiconductors are investigated. In the following chapter it will be shown how demanding the computation of thermodynamic properties from first-principles is with respect to both accuracy and run-time performance. Hence, we can combine the investigation of thermodynamic properties of III-V semiconductors with performing benchmarks of the S/PHI/nX program package. In order to evaluate the benchmark results, reference data have to be taken into account. In the DFT community one of the most important and widely applied codes is VASP [26], the Vienna ab-initio simulation package. This package became very successful due to its high accuracy and performance. It has been successfully employed to a wide range of applications. Therefore, it is a good choice to consider performance data obtained with the VASP[42] package as reference for the following benchmarks.

Both VASP and the current version of S/PHI/nX are plane-wave codes. VASP, however, employs ultrasoft pseudo potentials (USPP) and PAW while at the moment S/PHI/nX supports only norm-conserving pseudo potentials. For systems which require an accurate description of the semicore states, these states need to be treated as valence in the norm-conserving pseudo potential approach and a much higher energy-cutoff is required than a comparable simulation based on USPP. The goal of the benchmark is to verify that the S/PHI/nX ansatz yields a good run-time performance and not to confirm that USPP/PAW often outperforms norm-conserving pseudo potentials. Although the potentials of both packages differ important scaling information from such benchmarks can be derived:

---

[42] The tests have been performed with VASP 4.6 (serial version, pgf95) on AMD Opteron 246, 2.4 GHz.

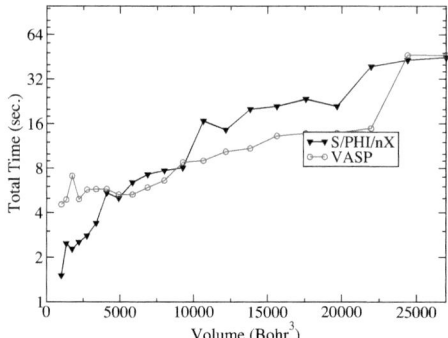

Figure 3.11: Run-time performance of S/PHI/nX and VASP in the low cut-off regime. Total execution time (user + system time) of a single Si atom in a simulation cell with varying volume. Both codes show the same overall scaling behavior with respect to the number of plane-waves.

1. The scaling behavior of both codes with respect to the unit cell volume or the number of plane-waves is expected to be similar since both are plane-wave codes. A worse scaling of S/PHI/nX would indicate problems of the abstract approach in S/PHI/nX. In contrast to conventional packages many operations which are usually manually written are automatically mapped in S/PHI/nX (pipelining: p. 58, BLAS/LAPACK mapping: p. 60, data type mapping: p. 64, Dirac projector mapping: p. 81). If the generic S/PHI/nX interface is not as efficient as human developers a drop in the scaling behavior schould be observed.

2. The comparison of the scaling behavior with respect to the number of atoms at the same energy cut-offs and FFT mesh sizes allows to test the efficiency of the non-local projectors.

3. Benchmarks with the optimal parameters for each code (energy cut-off, mesh sizes, etc.) can test the numerical accuracy that can be obtained with S/PHI/nX in comparison to VASP. Furthermore, the overall timing for realistic studies of systems can be compared.

The first test addresses the scaling behavior with respect to the unit cell volume by means of a single Si atom performed in a simulation cell with varying size. In this test the number of states and projectors remain constant while the number of plane-waves is being changed. This test has been performed with the energy cut-off $E_{\text{cut}}^{\text{S/PHI/nX,VASP}} = 10\,\text{Ry}$ and a $\mathbf{k}$ point at $(\frac{1}{4}\frac{1}{4}\frac{1}{4})$. The parameters to generate the pseudo potential are listed in the appendix. The results are presented in Fig. 3.11. The fluctuations of both data sets is due to the varying numbers of $\mathbf{G} + \mathbf{k}$ points at different volumes. S/PHI/nX and VASP show the same scaling behavior with respect to the system cell size. The execution speeds are comparable.

In Figs. 3.12 and 3.13 the run-time performance results for computing the total energy of cubic AlN bulk are depicted to analyze different aspects of the scaling behavior with respect to the number of atoms. As a measure, the time that is necessary to obtain the Born-Oppenheimer surface up to a reasonable accuracy of $\Delta E = 1 \cdot 10^{-8}\,\text{H}$ is taken to compute AlN with 2 atoms (1x1x1 fcc cell), 8 atoms (1x1x1 sc cell), 16 atoms (2x2x2 fcc cell), 54 atoms (3x3x3 fcc cell), 64 atoms (2x2x2 sc cell), 128 atoms (4x4x4 fcc cell), and 216 atoms (3x3x3 sc cell). This accuracy is required to obtain forces that are accurate enough to derive thermodynamic properties. This benchmark has been conducted with three settings: (a) identical energy

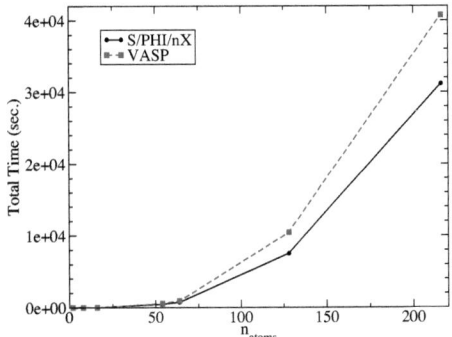

Figure 3.12: Scaling behavior of S/PHI/nX vs. VASP with respect to the number of atoms (AlN bulk). In order to compare the results, the calculations with both S/PHI/nX and VASP have been performed with the same settings for the energy cut-off $E_{\text{cut}} = 40\,\text{Ry}$ and the FFT meshes.

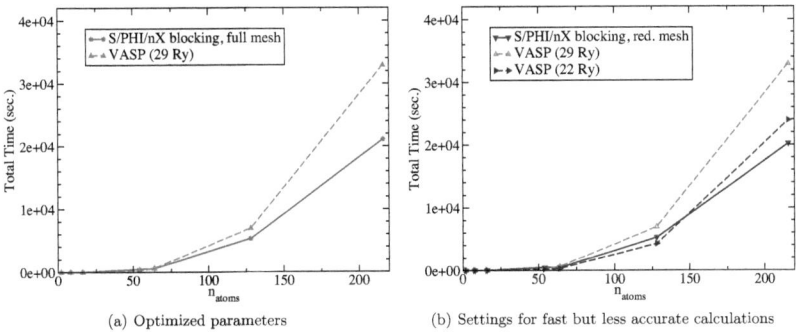

(a) Optimized parameters (b) Settings for fast but less accurate calculations

Figure 3.13: Performance comparison S/PHI/nX vs. VASP by means of AlN bulk using realistic parameter settings for both codes.

cut-offs and FFT mesh sizes, (b) typical settings for each code, and (c) optimized settings for each code to perform fast but less accurate calculations.

In Fig. 3.12 the benchmark results obtained with both codes and identical settings for E_{cut} and the FFT mesh size $n_x \times n_y \times n_z$ are displayed. It can be seen that both codes show similar results. The fact that S/PHI/nX is 25% faster than VASP to reach the given accuracy is likely due to the generalized eigenvalue problem of PAW which requires an additional computational effort. The test results indicate, that the caching technique which is necessary for applying the non-local projectors in S/PHI/nX are performing efficiently.

The previous benchmarks focused only on testing various technical aspects of the S/PHI/nX approach. From a user's point of view, however, only the actual execution speed is interesting. Therefore, in Fig. 3.13(a) the same set of calculations have been repeated with optimal settings for both codes, respectively. The VASP calculations have been performed with an optimal energy cut-off for this system of 29 Ry[43]. The FFT

---
[43]This value is suggested by the VASP potential file.

and augmentation meshes have been chosen to yield efficient and accurate results[44]. The norm-conserving pseudo potentials did not allow to decrease the energy cut-off below 40 Ry for the S/PHI/nX benchmark. In order to achieve optimal performance of S/PHI/nX for this test, full blocking (see Sec. 3.1.3 and 3.3.1) was enabled in the electronic minimization. It can be seen that S/PHI/nX is able to reach the same accuracy 30% faster than VASP in this example.

VASP provides the possibility to perform very fast but slightly less accurate calculations by reducing the energy cut-off to a minimum value and to use slightly under-sampled FFT meshes which leads then to minor wrap-around errors (see Sec. 2.1.1 on page 37). The benchmark results with reduced meshes are shown in Fig. 3.13(b). The VASP calculations have been performed with the minimum value of $E_{\text{cut}} = 22\,\text{Ry}$ while S/PHI/nX uses again 40 Ry. Both sets of calculations performed with S/PHI/nX and VASP applied the reduced FFT meshes. This test gives almost identical run-time performances of both codes.

In the last test we compute the thermodynamic properties $\alpha(T)$ and $C_{p,V}(T)$ of GaAs bulk as representative system with S/PHI/nX and compare the obtained results with VASP. The computed data are plotted in Fig. 3.14. The computational details (E_{cut}, **k**-point sampling, information about potential generation) can be found in the appendix. In the region of interest (T<300 K) both methods yield almost identical results. Only in the high temperature regime of $T > 1000\,\text{K}$ the slope of the linear expansion coefficients is slighly smaller ($\Delta \alpha \leq 2 \cdot 10^{-7} \text{K}^{-1}$) with our S/PHI/nX calculations. For this work we can, therefore, neglect the negative aspects of pseudoization. Note that with the same accuracy settings VASP's PAW calculation was 30% slower than the norm-conserving pseudopotential simulation performed with S/PHI/nX at a higher energy cut-off.

The performed benchmarks show that the highly abstract meta-language of S/PHI/nX which provides Dirac-notation, automatic blocking and memory management, can be used to develop a highly optimized and accurate DFT program package. In S/PHI/nX, the compiler "understands" the quantum mechanical context and can replace the required optimized algebraic operations with peak performance function calls. While in conventional program packages the human developer optimizes only some critical routines the S/PHI/nX approach shifts this task to the compiler and all routines are optimized. Optimization in S/PHI/nX focuses mainly on blocking techniques in order to exploit modern computer hardware. Since no or only little manual optimization is needed in S/PHI/nX the source code remains very short and intuitive.

It can be concluded that many tasks which had to be programmed manually can now be automatically handled during the compilation without loosing executional performance. While the developers can focus on implementation of quantum mechanical algorithms based on the S/PHI/nX Dirac notation the library performs tedious and error-prone tasks automatically. The above benchmarks show that S/PHI/nX is able to perform this mapping at least as efficiently as human developers.

3.5 Conclusions

In this chapter we discussed the ideas and concepts behind the S/PHI/nX density functional theory program package. Based on various new programming techniques an introduction of the Dirac notation to the C++ language became possible. By exploiting object-orientation we were able to mimic the building blocks of quantum mechanics in terms of C++ classes. The fundamental elements of this construction are Dirac vectors, Dirac basis-sets, Dirac wave functions. The hierarchy is "glued" with Dirac projectors. Since in

---
[44]using VASP parameters: ALGO=fast, ADDGRID=false, PREC=medium

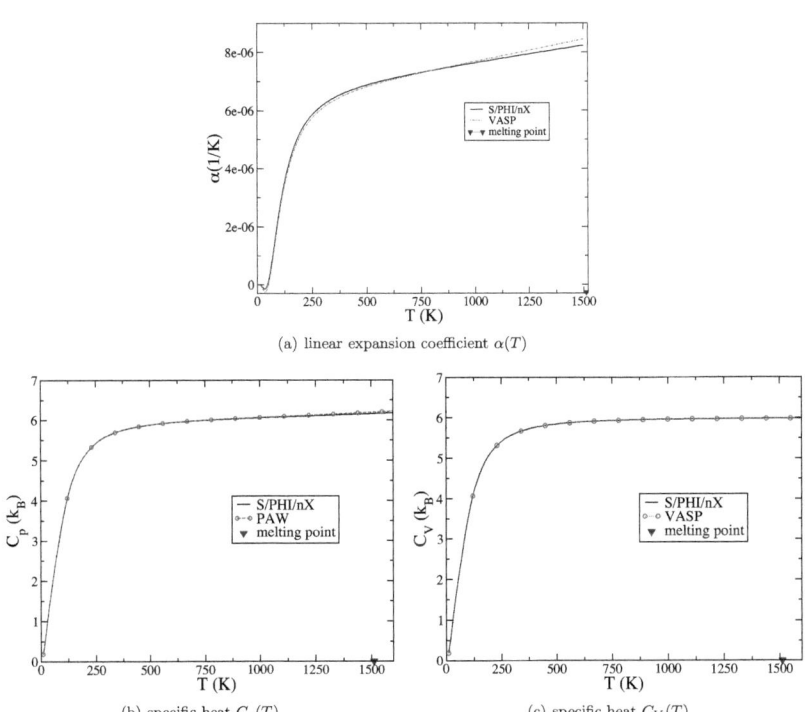

Figure 3.14: Comparison of the (a) linear expansion coefficient and (b) specific heat computed with norm-conserving pseudo potential plane-wave (PW-PS) with our S/PHI/nX code and with PAW using VASP. The results obtained with S/PHI/nX reflect almost perfectly the VASP results. Only in the high temperature regime ($T > 1000$ K) the data computed with S/PHI/nX and VASP differ slightly ($\Delta \alpha(T = 1500) = 2 \cdot 10^{-7}$ K^{-1} and $\Delta C_p(T = 1500) = 0.03\, k_B$).

the new ansatz the compiler is able to "understand" the quantum-mechanical context, it can replace the abstract Dirac terms with highly optimized numerical routines. In contrast to conventional programming, where human developers optimize only crucial parts of the program package, now the compiler performs an automatic code optimization throughout the entire package. Here, optimization refers to an automatic application of blocking algorithms and mapping to efficient function calls of high performance numeric libraries. It must not be confused with the intrinsic code optimization (e.g. "-O2"). The application of the Dirac notation in the source code has also significant advantages for the developer. Developing can now be accomplished in a physics language. Common issues of programming, such as memory management, calling unhandy functions from numeric libraries like BLAS or LAPACK are entirely shifted to the compiler, which speeds up the procedure of code developing, testing, and maintaining. The resulting code is dramatically shorter than usual packages. Using this approach we were able to derive the DFT Hamiltonian in only 550 code lines. The huge degree of flexibility could be combined with automatic peak performance. The entire library is prepared to work with other basis-sets.

Chapter 4

Applications

4.1 Introduction

In this section we apply the S/PHI/nX program package to compute thermodynamic properties such as the phonon dispersion curves $\omega_i(\mathbf{q})$, the linear expansion coefficient $\alpha(T)$, and the heat capacity $C_{p,V}(T)$ of III-V semiconductors. This semiconductor class is nowadays important for a manifold of applications. They play a major role when building up electronic and opto-electronic devices, including lasers, LEDs in the blue and UV regions of the spectra. It is important to understand their physical properties during the growth process as well as while they operate. Depending on the growth conditions their fabrication occurs at higher temperatures (e.g. growth of GaN at 950 K [117]), while they operate usually at room temperature.

As mentioned earlier (see p. 5) the derivation of trends is one of the fundamental tasks of CMD. In order to study thermodynamic trends in this work we investigate the most frequently applied III-V semiconducting systems which can be built up from combinations of neighboring elements in the periodic system: Al, Ga, In with N, P, and As, see Fig. 4.1. From these elements a matrix of the following 9 important semiconductors can be systematically investigated: AlAs, GaAs, InAs, AlP, GaP, InP, AlN, GaN, and InN which can act as a basis for deriving important thermodynamical trends. These systems crystallize in the wurzite and/or zincblende phase. In the zincblende phase many of them[1] are reported to exhibit a thermal expansion anomaly, i.e., up to a critical temperature the lattice parameter decreases with increasing temperatures.

---
[1] For the nitrides in the zincblende phase no experimental data for $\alpha(T)$ were found.

Figure 4.1: Location in the periodic systems and electronic configuration of the involved elements to investigate the following systems: AlAs, GaAs, InAs, AlP, GaP, InP, AlN, GaN, InN.

From the technological point of view the exact knowledge of this anomaly is important since it determines, e.g., the thermal lattice mismatch between substrate and semiconductor. It will be shown in the following that for most of the above mentioned systems the experimentally available data on the linear expansion coefficients scatter significantly (in particular for AlAs, InAs, GaP and InP).

Although in literature thermodynamic properties have been calculated from first-principles there is only little known about the influence of the XC functional. The previously performed theoretical studies applied mainly LDA. In this chapter we present our results with both the LDA and the GGA-PBE functional in order to roughly estimate how the choice of the XC functional influences the accuracy of the computed $\alpha(T)$ and $C_{p,V}(T)$.

The details about the applied pseudo potentials are given in the appendix. For Ga and In the semicore d-states can be treated either as core or as valence states in the pseudo potential approach. In Refs. [118, 119] it has been shown that an explicit treatment of the d-semicore states in the valence improves the structural and cohesive properties (such as a_0 and E_b) significantly. Therefore, we expect that the choice of the treatment of the d electron influences also the description of ω, $\alpha(T)$, and $C_{p,V}(T)$. Following Ref. [119] we constructed the pseudo potentials Ga^{NLCC}, In^{NLCC} which employ non-linear core correction [120] as well as the pseudo potentials Ga^{3d}, In^{4d} which include the 3d/4d electrons in the valence part. Following Ref. [119] for these two systems we also apply non-local projectors for the s, p, d components, and the f component as the local potential.

4.2 Thermodynamic properties

4.2.1 Convergence aspects

The derivative nature of the thermodynamical properties suggests a severe effect of convergence issues on the quality of properties computed from first-principles: $C_p(T)$, $C_V(T)$, and $\gamma(T)$ can be expressed as second derivatives of the free energy $F(T,V)$ (see Eqs. (2.81), (2.82), and (2.83)) which in turn depends on the forces via Eqs. (2.70), (2.71), and (2.75). As introduced in Sec. 1.7.1 forces are derivatives of the total energy surface. Hence, very small numerical errors in the total energy arising from, e.g., an incompleteness of the basis-set, k-point sampling, or small problems related to the pseudo potentials, will largely influence the accuracy of the obtained free energy surface. This has been observed for metallic systems [37].

Thus, we first focus on convergence aspects with respect to E_{cut} and the k-point mesh density. Furthermore, we estimate the influence of the pseudo potential approach for the achievable accuracy of the calculated thermodynamic properties in comparison to a PAW description. We demonstrate the general approach by means of the GaAs bulk system as a representative. For the other systems analogous tests have been conducted. The obtained converged parameters for the investigated systems are presented in the appendix.

Convergence of bulk properties at T=0 K

A high accuracy of the equilibrium lattice constant a_0, the Bulk modulus B and its derivative B' at T=0 K is crucial as these data provide the reference values for the subsequent temperature-dependent entities. In Tab. 4.1 they are presented for various energy cut-offs. In order to estimate the intrinsic uncertainty of DFT that exists due to the treatment of the exchange-correlation potential for the investigated systems, we have calculated the entities with both LDA and PBE. The comparison with FP-LAPW calculations allows a rough estimation of the error introduced in the frozen-core approximation and the pseudoization procedure. Above an energy cut-off of 15 Ry the computed data show excellent agreement ($\Delta a \leq 0.04$ Å) with FP-LAPW data [121] as well as the experiment [122, 123, 124]. Similarly to many other systems LDA tends also for GaAs bulk to overbind while PBE underbinds slightly.

E_{cut} (Ry)	a (Å)	a/a_0 (%)	a_0/a_0^{exp} (%)	B (GPa)	B/B_0 (%)	B/B^{exp} (%)	B'
(a) Experiment							
	5.65 [122]			76.9 [123]			4.80 [124]
(b) LDA							
10	5.794	+3.07	+2.49	103.28	+29.04	+25.55	0.99
15	5.631	+0.27	-0.34	75.177	+2.50	-2.29	5.12
20	5.622	+0.11	-0.50	72.05	-1.72	-6.72	4.44
25	5.616	+0.00	-0.61	73.88	+0.79	-4.08	4.21
30	5.615	-0.02	-0.62	73.39	+0.13	-4.78	4.36
40	5.615	-0.02	-0.62	74.24	+1.28	-3.57	4.17
50	5.616	+0.00	-0.62	73.29	+0.00	-4.92	4.29
FP-LAPW	5.621 [121]			74.2 [121]			
(c) PBE							
10	6.008	+4.20	+6.34	92.02	+59.97	+19.67	15.41
15	5.779	+0.23	+2.28	56.42	-1.92	-26.63	5.31
20	5.776	+0.17	+2.23	59.09	+2.72	-23.16	4.25
25	5.768	+0.03	+2.09	58.12	+1.04	-24.41	4.51
30	5.766	+0.00	+2.05	57.66	+0.25	-25.01	4.67
40	5.766	+0.00	+2.05	57.52	+0.00	-25.19	4.68
FP-LAPW	5.74 [125]			59.96 [125]			

Table 4.1: Convergence of GaAs bulk properties with resp. to the energy cut-off E_{cut} in comparison with (a) the experiment, (b) **LDA**, and (c) **PBE**: The computed equilibrium lattice constant a and the bulk modulus B are compared with the converged reference values at 50 Ry (a_0 and B_0) and the experimental values. Both have been extrapolated to T=0 K. B' denotes the computed bulk modulus derivative.

Convergence of thermodynamic properties

Based on the computed T=0 K values from the previous subsection the temperature dependence of both the linear expansion coefficients and the heat capacity can be investigated at various energy-cutoff values and **k**-point meshes.

When computing the phonon spectra in the direct approach the simulation cells have to be chosen large enough to properly describe long range spatial interactions, i.e., the phonon branches close to Γ. The second effect of the cell size is the sampling of the Brillouin zone with exact **q**-points. For example, by doubling the unit cell **q**-points such as X can be sampled exactly[2]. If the simulation cell is too small, an unphysical

---
[2]consider a frozen phonon calculation of a linear atomic chain: the phonon at **X** can be described by constructing a super cell obtained by repeating the unit cell twice along the x axis and displacing the atom (0,0,0) to ($\Delta x, 0, 0$).

interaction between the displaced atom and the mirror atoms in the periodically repeated unit cells occurs. The forces acting on the displaced atoms are artificially screened by those from the displaced image atoms. This effect reduces the corresponding phonon frequencies. Only if the cell is sufficiently large the artificial forces originating from the displaced image atoms are screened by the undisplaced atoms and the phonon frequencies can be obtained correctly. This effect is illustrated in Fig. 4.3. Using a cell of 64 atoms (2x2x2 folding of the conventional cubic cell) a *minor* reduction of the phonon frequencies can be observed at $\frac{1}{4}$K which disappears already at $\frac{1}{2}$K. The artificial red shift is for all systems investigated here less than 1 meV and contributes only to a very small part of the **q**-space. The effect of this *minor* softening at $\frac{1}{4}$K is estimated to be negligible when integrating over the entire Brillouin zone and does, therefore, not justify the significantly larger computational costs.

In Fig. 4.4 the influence of the plane-wave energy cut-off on $\alpha(T)$ and $C_p(T)$ is being shown. The heat capacity (Fig. 4.4b) is surprisingly insensitive compared to the linear expansion coefficient (Fig. 4.4(a)). A value of about 15 Ry yields already converged results ($\Delta C_p < 0.1\,\mathrm{J\,mol^{-1}K^{-1}}$).

In case of the linear expansion coefficients $\alpha(T)$, however, this picture changes. At 15 Ry the differences to the 35 Ry curve is still significant. In order to obtain a converged *absolute* value of $\alpha(T = 500\,\mathrm{K})$ an energy cut-off of at least 25 Ry is necessary. That is surprising since according to Eq. (2.80) such a sensitivity would not be expected compared to $C_p(T)$. However, while the *absolute* values of the linear expansion coefficients are still varying, at $E_\mathrm{cut} > 15\,\mathrm{Ry}$ the *slope* of $\alpha(T)$ is already converged. The shift of the *absolute* values can be explained by means of the abnormal thermal expansion behavior of GaAs at low temperatures. In the regime at about $T < 80\,\mathrm{K}$ GaAs shows an abnormal thermal compression. With increasing temperatures the volume contracts and $\alpha(T) < 0$. Only at approx. $T > 80\,\mathrm{K}$ GaAs shows the usual expansion behavior. This well-known behavior (see for example Ref. [40]) of the III-V semiconductors in the zincblende phase can be explained by means of the mode-Grüneisen parameters $\gamma(\mathbf{k})$. According to Ref. [126] (Fig. 4.2(b and c)) the energetically lowest transversal-acoustic phonon branches[3] TA, TA1, and TA2 at Γ, L and X have all negative mode-Grüneisen parameters. At low temperatures mainly these negative branches are occupied. With $\gamma < 0$ the entropy decreases with increasing lattice constant and $-S^\mathrm{vib}T$ shows a positive slope [127]. Consequently, the lattice contracts compared to the T=0 K lattice constant. With rising temperatures also energetically higher lying phonon states with positive γ values become occupied, which drives a normal expansion behavior.

Both γ and α depend via Eqs. (2.77), (2.76), (2.83), and (2.80) on the volume dependence $d\omega_i/dV$ which is due to its derivative nature numerically sensitive. Small numerical inaccuracies can induce inaccuracies in γ which consequently introduce inaccuracies in the prediction of the expansion behavior (Eq. 2.80). An accurate description of the location and magnitude of the thermal anomaly requires thus a high accuracy of $d\omega_i/dV$ as can be seen in Fig. 4.4a. In contrast to the discussion above (Tab. 4.1) here an energy cut-off of 25 Ry is necessary to obtain the location and magnitude of the thermal anomaly within acceptable accuracy ($\Delta\alpha < 1\cdot 10^{-7}\,\mathrm{K^{-1}}$). This can be generalized to all systems with the same expansion behavior: a poor description of the low temperature limit introduces a shift in the absolute values at high temperatures for systems with such an abnormal thermal expansion behavior. This is due to a poor sampling of the volume dependence of ω_i and, therefore, the mode-Grüneisen parameters $\gamma(\mathbf{k})$.

The analysis of the influence of the energy cut-off to $\alpha(T)$ and $C_p(T)$ provides already a rough estimation of the required energy cut-off based on the (computationally inexpensive) T=0 K bulk properties. At the minimum cut-off of 25 Ry the equilibrium lattice constant is already converged to $\Delta a < 0.001\,\mathrm{\AA}$ and

---

[3]The nomenclature of phonon branches we use throughout this work can be seen in Fig. 4.2(a).

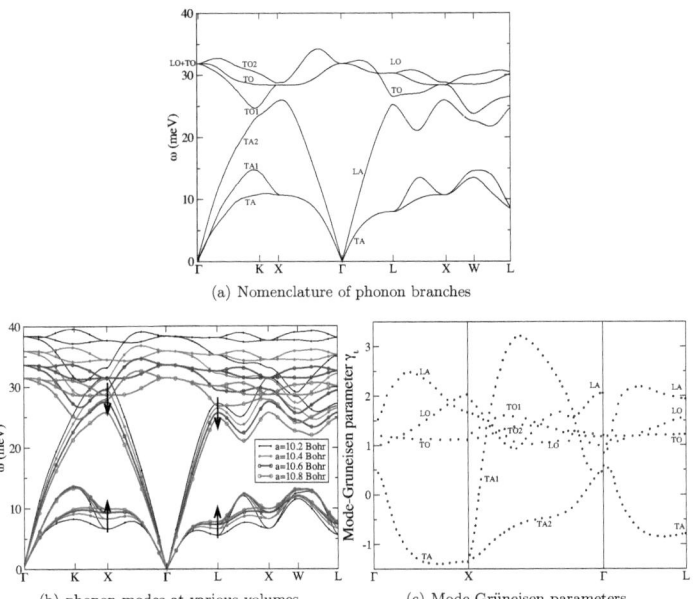

Figure 4.2: Phonon modes that occur in zincblende semiconductors, shown by means of GaAs bulk. (a) Nomenclature of phonon branches. The energetically lowest states are transversal acoustic branches (TA1 and TA2) which become degenerate at X. The transversal optical phonon (TO) is split into TO1 and TO2. (b) Volume dependence of $\omega_i(\mathbf{q})$. The phonon frequencies of $\omega_{TA}(X)$ and $\omega_{TA2}(L)$ increase with increasing lattice constant. (c) Mode-Grüneisen parameters $\gamma(\mathbf{q})$ from Ref. [126]. Most of the III-V and II-VI semiconductors show negative thermal expansion coefficients at low temperatures [40]. That is due to the (flat) TA phonon modes possessing negative mode-Grüneisen parameters.

$\Delta B < 1\%$ as shown in Tab. 4.1. The first value is crucial when determining α, while the phonon frequencies are correlated with the bulk modulus B.

In analogy to the above considerations, in Fig. 4.5 the influence of the sampling of the Brillouin-zone on $\alpha(T)$ and $C_p(T)$ is depicted. As for the energy cut-off also the number of $\mathbf{k}$-points has a stronger influence on the absolute values of $\alpha(T)$ than to those of $C_p(T)$. A sampling of $3 \times 3 \times 3$ $\mathbf{k}$-points yields converged data with an accuracy of $\Delta \alpha < 1 \cdot 10^{-8} \, \mathrm{K}^{-1}$ and $\Delta C_p < 0.1 \, \mathrm{Jmol}^{-1}\mathrm{K}^{-1}$.

4.3 Comparison with experiment

4.3.1 Bulk properties at T=0 K

In order to evaluate the free energy surface at T=0 K the structural bulk properties a_0, B, and B' have been computed using the Murnaghan equation of state Eq. (2.76). In Tab. 4.2 the experimental data are compared with our results for all 9 considered semiconductors obtained from LDA and PBE calculations. For the investigated **arsenides** the equilibrium lattice constants are in good agreement with the experimental

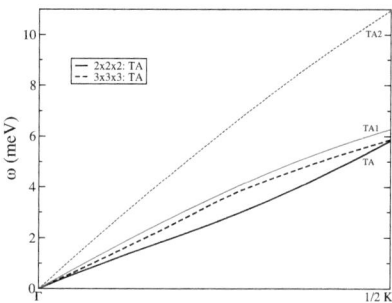

Figure 4.3: Influence of the unit cell size to the long range limit of the acoustic phonon branches TA, TA1, and TA2 in the interval $[\Gamma, \frac{1}{2}K]$. The TA branch of GaAs bulk computed with a 64 atom (2x2x2) unit cell (bold solid curve) shows a minor softening which disappears for the larger simulation 216 atom cells (bold dashed curve).

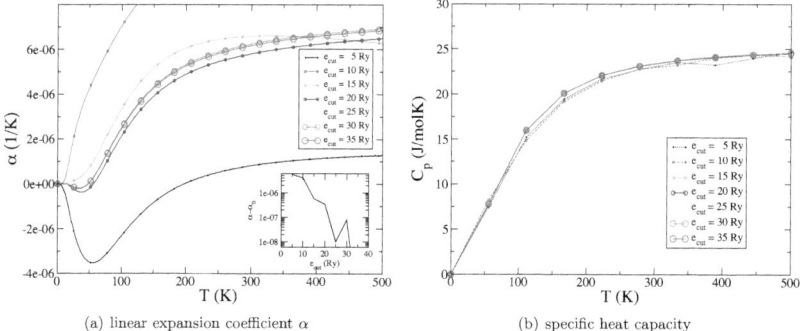

(a) linear expansion coefficient α (b) specific heat capacity

Figure 4.4: Convergence of (a) the linear expansion coefficient and (b) the specific heat of GaAs with respect to the energy cut-off. The tests have been performed on a **k**-point mesh of 3x3x3. The inset in Fig. (a) shows the convergence of α with respect to the converged value α_0 at 35 Ry.

values. Like for many other materials also the lattice constants of the arsenides obtained from LDA follow the general trend of underestimating the lattice constant slightly while PBE shows the opposite behavior. The differences to the experimental values are always well below 2.5%. The bulk moduli and their derivatives are also described well: the deviations to the experimentally obtained bulk moduli for AlAs and InAs are less than 16%. Only in case of GaAs bulk the PBE value is underestimated by about 25%. However, this is in good agreement with other norm-conserving GGA pseudo potential simulations: in Ref. [140, 141] the influence of the chosen GGA functional to the deviation of B is investigated. In their work a value of 56 GPa has been obtained which is close to our PBE-value of 57.2 GPa.

In case of the **phospides** (second row of Tab. 4.2) the prediction of the lattice constant at T=0 K is even closer to experiment. All deviations are less than 1.7%. As for the arsenides also in case of the phosphides LDA(PBE) tends to over(under)bind slightly. We obtained bulk moduli with deviations to the experiment of less than 10%. Please note, that for AlP (zincblende) no experimental data for the bulk modulus was found. Therefore we compare our AlP results with FP-LMTO calculations.

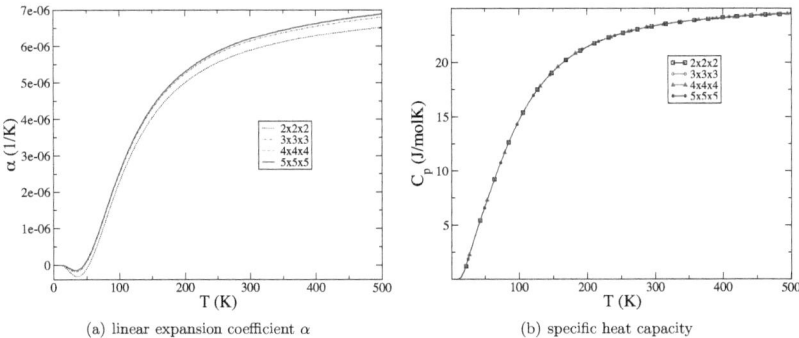

Figure 4.5: Convergence of (a) the linear expansion coefficient and (b) the specific heat of GaAs with respect to the **k**-point density. As expected for a semiconductor only a few **k** points are required to obtain an accurate Brillouin zone integration. The calculations have been performed with an energy cut-off of 25 Ry. The linear expansion coefficient requires a Monkhorst Pack folding of at least 3x3x3 while the specific heat is already well converged with a sampling of 2x2x2.

In the last two rows of Tab. 4.2 our LDA and PBE results for the **nitrides** have been compared with the experiment or other theoretical calculations. The computed lattice constants of AlN and GaN-NLCC are close to the experiment with deviations less than 1.3%. As for the previously discussed systems also here LDA(PBE) under(over)estimates the lattice constant compared to the experimentally obtained values. In InN-NLCC, however, the picture is different since both LDA and PBE underestimate the lattice constant. The poor description of InN using NLCC is well known (see, e.g., Ref. [119]). InN is also particularly sensitive to the choice of the description of the exchange-correlation potential. For example, the band gap is reported to be positive ($E_g = 0.16$eV) with LDA/NLCC [142] and negative with LDA/4d [118] ($E_g^{\text{LDA},4d} = -0.40$eV) as well as PBE/4d [118] ($E_g^{\text{PBE},4d} = -0.55$eV).

Conclusion

For all investigated systems the convergence parameters that are necessary to obtain accurate thermodynamical properties have been determined carefully. In order to determine the influence of the exchange-correlation potentials the convergence analysis has been performed with both LDA and PBE. The general trend of over(under)estimating the equilibrium lattice constant could be confirmed to be valid for the investigated systems. With the determined convergence parameters for E_{cut} and the **k**-point sampling all computed lattice constants are in good agreement with the experiment. The deviations are always less than 2.1%. Only the predicted bulk modulus obtained with the PBE functional deviates in case of GaAs and InN by 25% and 18%, respectively.

4.3.2 Phonon spectra

In this section the computed phonon spectra are presented which are needed to derive the thermodynamic properties. A particular focus will be on of the exchange-correlation functional as well as the treatment of the semicore d-states of Ga and In is investigated. In order to compare our results with the literature the phonon dispersion curves have been calculated at the temperatures of the experiment. To do so, the corresponding

	AlAs			GaAs			InAs		
	LDA	Exp.	PBE	LDA	Exp.	PBE	LDA	Exp.	PBE
a	5.628	5.663 [128]	5.749	5.616	5.65 [122]	5.766	5.978	6.058 [129]	6.131
Δa	-0.62		+1.50	-0.62		+2.05	-1.34		+1.19
B	73.10	74.4 [130]	65.28	73.29	76.9 [123]	57.52	63.12	59.2 [131]	51.04
ΔB	-1.78		-13.96	-4.92		-25.19	+6.22		-15.98
B'	4.27	5.0 [130]	4.03	4.29	4.8 [124]	4.68	4.90	6.8[131]	4.50
	AlP			GaP			InP		
	LDA	Exp.	PBE	LDA	Exp.	PBE	LDA	Exp.	PBE
a	5.413	5.451 [132]	5.452	5.398	5.447 [42]	5.511	5.772	5.866 [40]	5.886
Δa	-0.99		+0.00	-0.91		+1.16	-1.63		+0.34
B	88.41	87* [133]	84.98	90.40	87.4 [134]	80.45	77.45	76 [135]	65.39
ΔB	+1.59		-2.37	+3.32		-8.63	+1.88		-8.57
B'	3.65	4.30 [133]	3.61	3.98	4.5 [136, 134]	3.78	4.76	4.0 [135]	4.52
	AlN			GaN-NLCC			InN-NLCC		
	LDA	Exp.	PBE	LDA	Exp.	PBE	LDA	Exp.	PBE
a	4.03	4.36 [137]	4.387	4.449	4.5 [138]	4.525	4.83	4.98 [137]	4.92
Δa	-1.32		+0.62	-1.15		+0.55	-3.11		-1.16
B	206.84	202 [137]	191.15	214.02	190 [139]	177.02	182.51	137 [139]	146.64
ΔB	+2.34		-5.68	+11.22		-7.33	+24.94		+6.59
B'	3.77	4.15$^\text{G}$ [138]	3.89	5.12	4.27$^\text{G}$ [138]	4.69	3.90	4.43$^\text{G}$ [118]	4.1
				GaN-3d			InN-4d		
				LDA	Exp.	PBE	LDA	Exp.	PBE
a				4.436	4.5 [138]	4.5	4.97	4.98 [137]	5.1
Δa				-1.44		0	-0.04		+2.1
B				213.42	190 [139]	195.31	136.79	137 [139]	116.18
ΔB				+10.98		+2.72	-0.02		-17.9
B'				5.12	4.27$^\text{G}$ [138]	4.92	4.52	4.43$^\text{G}$ [118]	4.24

Table 4.2: Computed bulk properties of all investigated systems with LDA and PBE. Lattice constants a given in Å, Bulk modulus B in GPa, and deviations to reference data Δa and ΔB in %. For systems where experimental data are not available, other theoretical data act as a reference: Values labeled with (*) refer to FP-LMTO-LDA calculation and $^\text{G}$ marks pseudo potential plane-wave GGA results rather than experimental ones.

volumes $V = V_\text{exp}$ have been extrapolated consistently from our LDA/PBE thermal expansion computations (Eqs. (2.79), (2.80)). The phonon dispersion curves $\omega_i(\mathbf{q})|_{V_\text{exp}}$ have been linearly interpolated [92] from dispersion curves at different volumes $\omega_i(\mathbf{q})_{V=V_0 \pm 3\%}$. In case of structures for which experimental data are not available we overlayed our spectra with other theoretical data, which are usually performed without including temperature effects. In this case we calculate our spectra also at T=0 K, i.e., at the equilibrium lattice constants of LDA and PBE, respectively.

Arsenides In Fig. 4.6 the phonon dispersion curves of the arsenides are presented. Similar to the study of Grabowski and co-workers [37] for metals, the frequencies obtained from LDA(PBE) under(over)estimate the experimental data. As discussed earlier our data show a *minor* red shift of the acoustic branches in the long-range limit due to the relatively small 64 atom cell. From Fig. 4.6 it can be seen that there is a significant discrepancy between our results and experimental data in the optical branches in the vicinity of the Γ point. The experiment predicts a splitting of the LO and TO phonon branches while our data displays a degeneracy. This difference can be explained with the missing Born effective charge tensor in our calculations (see p. 50).

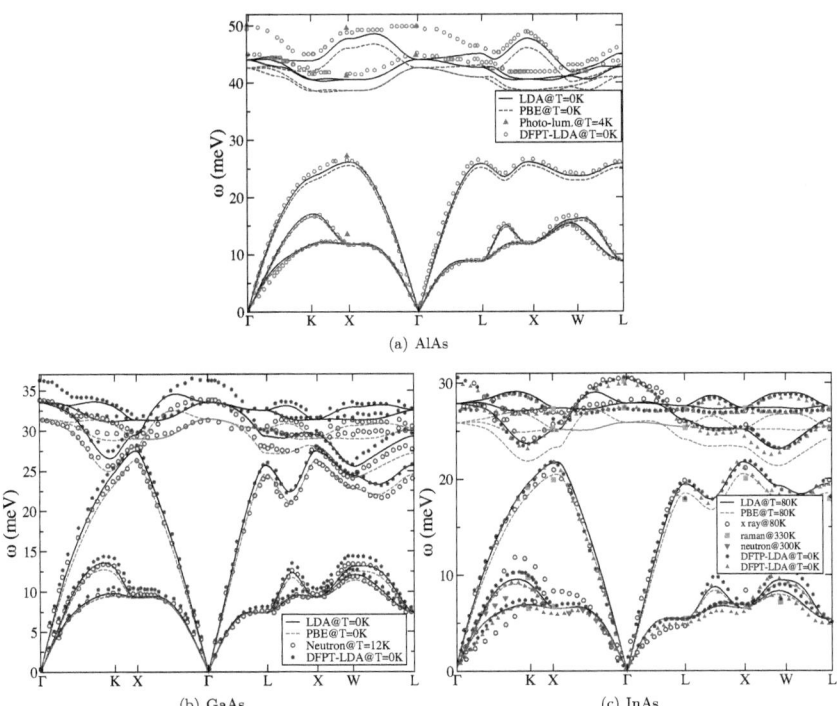

Figure 4.6: Phonon spectra calculated at the theoretical equilibrium lattice constants of LDA and PBE compared with the experimental data. (a) **AlAs**. triangles: Raman scattering data at T=4 K [143, 144], circles: pseudo potential DFPT-LDA [108]. (b) **GaAs**. open circles: experimental low temperature neutron scattering data [145] at T=12 K, closed circles: DFPT-LDA [108] (c) **InAs**. open circles: therm. diffuse X-ray data at T=80 K[146], boxes: Raman data at room temperature [147], triangles: neutron scattering data at room temperature [146] , DFPT-LDA taken from Ref. [148] (solid circles) and from Ref. [149] (triangles up).

For **AlAs** only experimental data of the phonon spectra at X and Γ are available [143, 144]. We added also a computed spectrum obtained in another study employing pseudo potentials [108]. The experimental data points have been obtained by low temperature Raman scattering techniques at 4 K while Giannozzi *et al.* performed LDA calculations based on Density Functional Perturbation Theory (DFPT) for the T=0 K case.

Our results for the energetically lower lying acoustic branches, also obtained at T=0 K, are virtually identical to those obtained by the linear response approach employed by Giannozi. Both theoretical data (DFPT-LDA and ours) are approx. 2 meV smaller at X compared to the experimental value. Minor differences can also be seen in the TA branch between Γ and K which is related to the finite size of our simulation cell. The position of the optical branches is slightly underestimated by approx. 1 meV in our calculations compared to the DFPT-LDA and Raman data.

For the well studied system of **GaAs** among others Strauch [145] presented phonon spectra based on the inelastic neutron scattering technique at low temperatures (T=12 K). In order to compare our results with additional reference data we also added theoretical data based on the linear response method [108] using LDA as exchange-correlation functional. For all acoustic phonons the experimental data are very close to our computed LDA and PBE phonon spectra. The DFPT-LDA results are, however, by approx. 1 meV larger than our LDA data as well as the experiment. This deviation is most probably due to different pseudo potentials. The high-symmetry path $\Gamma \rightarrow$ K is, in contrast to AlAs, almost the same as the experiment and the linear response data. There is a margin of error of about 3 meV between experiment, the DFPT-LDA and our data which is due to the lack of the LO-TO splitting in our approach.

The large scattering of the experimental data for **InAs** does not allow for a direct verification of our theoretical results. Since we did not find experimental data at low temperatures we extrapolated (see p. 115) our LDA and PBE results to 80 K which corresponds to the conditions of the X-ray investigations [146]. Again a discrepancy at $\Gamma \rightarrow$ K can be seen which, however, agrees with other theoretical studies ([148], [149]).

Phosphides The phonon spectra of the phosphides are shown in Fig. 4.7. The general shape of the phonon spectra is similar to that of the investigated arsenides in the zincblende structure due to the same symmetries (space group F$\bar{4}$3m).

In Fig. 4.7(a) we compare our LDA and PBE results for **AlP** with low temperature (5 K) Raman data [143] at the Γ-point as well as theoretical phonons spectra obtained from DFPT-LDA pseudo potential investigations [126]. It can be seen that the influence of the chosen exchange-correlation functional to the phonon spectra is very small for the zincblende AlP system. The DFPT data are virtually the same as ours, only the long wavelength limit is slightly softer than ours which is due to the finite simulation cell size in our approach. Also the optical branches are in very good agreement with both data sets, except the expected lack of the splitting at the Γ-point.

Similarly good results could be obtained for the **GaP** system. Our data agree very well with the experimentally obtained neutron spectra [150, 153] as well as the previously reported DFT-LDA spectra[148]. Only at the L point our frequencies are by about 2 meV larger than the experimentally obtained frequencies. Our data are, however, very close with the DFT-LDA curves from Ref. [148].

For the **InP** at low temperatures only Raman data at Γ obtained at T=4 K are available [154]. The experimental TO data are in good agreement with our results as can be seen in Fig. 4.7(c). For low temperatures there are no experimental data available. We added results obtained from DFPT-LDA calculations [155] which are up to 1 meV close to ours. In the long wavelength limit our phonon branches appear slightly harder. Besides the T=4 K Raman data there are also coherent inelastic neutron scattering experimental data [153] available at room temperature. We have, therefore, evaluated the corresponding theoretical phonon spectra at T=300 K. Our results indicate that an increase of the temperature leads to a slight red shift of the phonon branches. The calculated shift is less than 0.5 meV for the acoustic and 1.2 meV for the optical phonons.

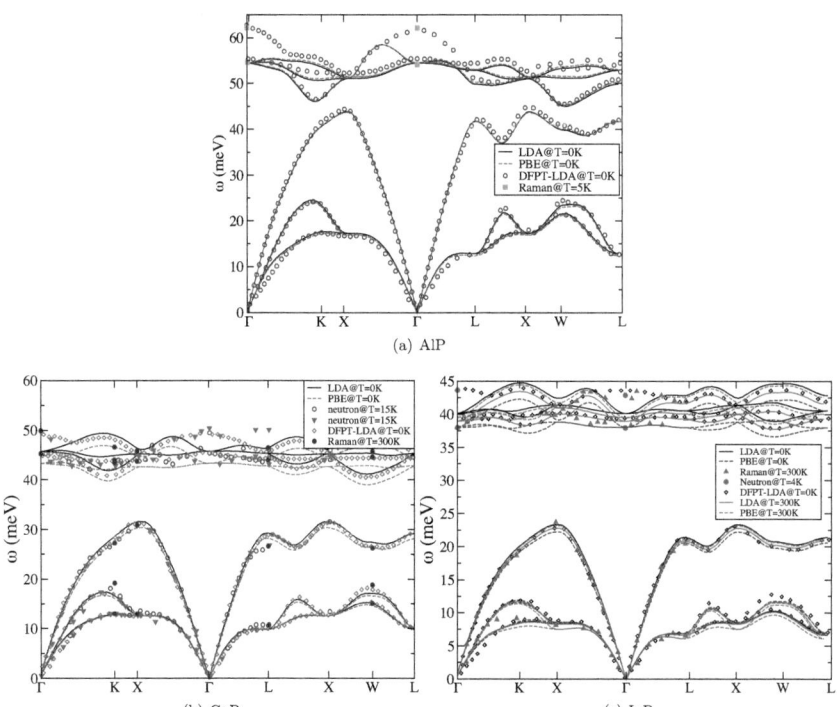

Figure 4.7: Phonon spectra calculated at the theoretical equilibrium constants of LDA and PBE compared with the experimental data. (a) **AlP**. solid boxes: low temperature Raman data [143], circles: DFPT-LDA [126] (b) **GaP**. Neutron data at 15 K from [150](circles) and [151] (triangles down), room temperature Raman data [152] (solid circles), DFPT-LDA [148] (diamonds) (c) **InP**. solid circles: low temperature neutron data [153], triangles: room temperature Raman data [154], diamonds: DFPT-LDA [155]

Nitrides Since no experimental data are available for cubic **AlN**, in Fig. 4.8(a) we compare our results with another DFT-LDA investigation at T=0 K [156, 157]. Except the missing LO-TO splitting as well as a slightly harder phonon in the long wavelength limit our data are in good agreement with the other theoretical work.

In Fig. 4.8(b)-(c) we show the computed **GaN** phonon spectra evaluated employing with either NLCC or an explicit description of the 3d electrons as valence (see p. 110). This figure also displays the available experimental (low temperature Raman [158]) and theoretical (DFPT-LDA [156, 157]) data.

When the semicore states of Ga are treated as core states and taking the non-linear core correction (NLCC) into account, the bonds appear stronger and the bond distances shorter than all-electron calculations would suggest (see Ref. [119]). Our data are consistent with this trend: The stronger bonds obtained with LDA-NLCC and PBE-NLCC result in higher phonon frequencies in comparison to those obtained with LDA-3d or PBE-3d respectively. There is an increasing red shift of the phonon frequencies for different functionals in the order LDA-NLCC, LDA-3d, PBE-NLCC, PBE-3d. This behavior can be assigned to the description

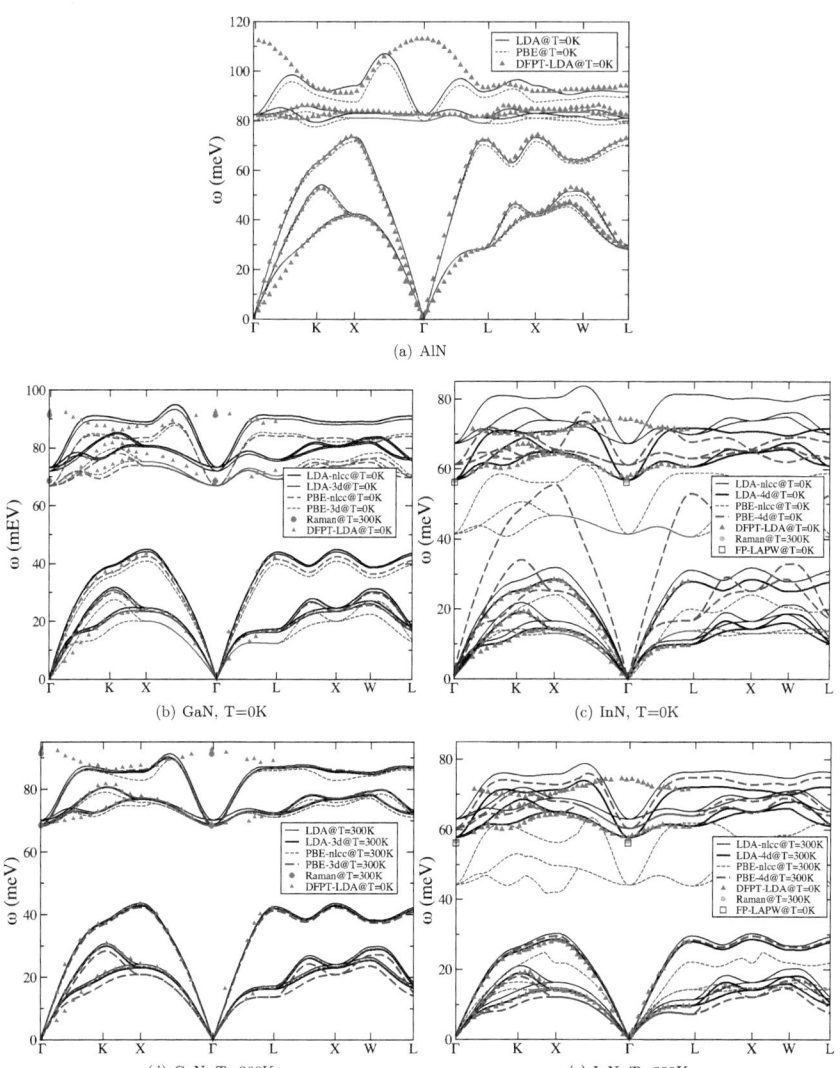

Figure 4.8: Phonon spectra at the theoretical equilibrium constants of LDA and PBE compared with the experimental data. (g) **AlN**. *ab-initio* data (theory) from [156, 157], (h) **GaN**. exp. data from [158], *ab-initio* data (theory) from [156, 157], (i) **InN**. *ab-initio* data (theory) from [156, 157]

of the band gap with the different exchange-correlation functionals. In Fig. 4.9(b) we show the computed electronic band structures of GaN with LDA-3d/NLCC and PBE-3d/NLCC. Our PBE pseudo potentials yield one-particle energies $\varepsilon_i(\mathbf{k})$ very close to those obtained with FP-LAPW-GGA [159]. That indicates that the pseudoization has no negative influence to the quality of our obtained electronic band structures. The experimental band gap is reported to be 3.23 eV [160]. Our computed band gaps are 2.18 eV (LDA-NLCC), 1.88 eV (LDA-3d), 1.81 eV (PBE-NLCC), and 1.66 eV (PBE-3d). They are close to results obtained elsewhere (2.20 eV LDA-NLCC [161], 1.89 eV LDA-3d [118], 1.99 eV PBE-NLCC [162], 1.74 PBE-3d [159]). By formally "improving" the exchange-correlation potential from LDA to PBE and by describing the d-electrons as valence, the computed band gap deviates further away from the experimental value. This artificially decreased band gap changes the electronic behavior of the described GaN system to be more metal-like, i.e., an artificial metallic-like screening effect occurs. The screening reduces the interactions between the atoms and thus, reduces the phonon frequencies. Therefore, the red shift of the phonon frequencies correlates with the corresponding band gap (see inset of Fig.4.9a).

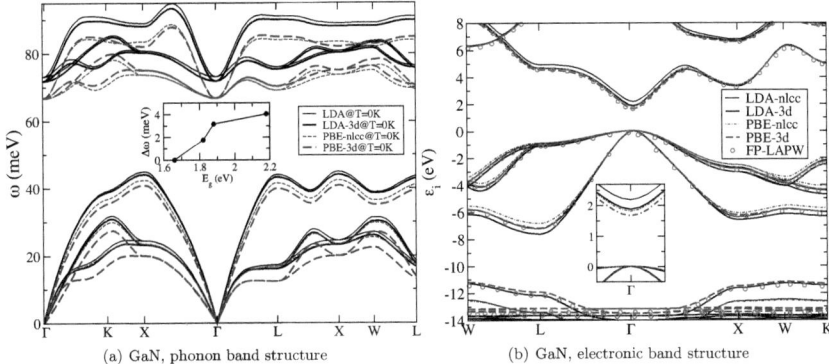

(a) GaN, phonon band structure   (b) GaN, electronic band structure

Figure 4.9: (a) Phonon dispersion curves as in Fig. 4.8b and (b) electronic band structure of GaN at T=0 K at the theoretical lattice constants. The top of the valence band has been aligned to the Fermi level (0 eV). The inset of figure (b) magnifies the area around the direct band gap at the Γ point. Beside the computed dispersion curves for LDA-NLCC, LDA-3d, PBE-NLCC, and PBE-3d results obtained from FP-LAPW [159] using PBE have been added. The shift the phonon frequency $\Delta\omega_{TA}(X)$ vs. the band gap E_g is shown in the inset of (a).

In Fig. 4.8(d)-(e) our **InN** results are presented. Since the experimentally obtained Raman spectra have been measured at room temperature [163] we extrapolated our phonon spectra to the same conditions (T=300 K) and find a deviation of less than 3 meV. Since the experiment provides only a value at the Γ-point, we added a DFPT-LDA data set [164] as well as results obtained from FP-LAPW [163]. It can be seen that our LDA calculation with the 4d valence electrons yields almost the same result as that taken from Ref. [164]. Our LDA-NLCC and LDA-4d results are in good agreement with both experimental and other theoretical findings and suggest a good basis for the subsequent thermodynamic calculations. While the LDA results yield accurate phonon spectra the predicted phonon frequencies obtained with PBE are significantly overestimated. This is likely related to the PBE functional itself. PBE is known to predict a number of properties for the InN system poorly. Fuchs et al. [119] reported even a negative formation enthalpy, i.e., an endothermic behavior of InN ($\Delta H^{exp} = -0.18\,\text{eV}$, $\Delta H^{LDA} = -0.19\,\text{eV}$, $\Delta H^{PBE} = +0.35\,\text{eV}$). This well-known shortcoming of PBE in case of zincblende InN leads apparently also to a poor description of the phonon spectra. InN is also known to be challenging when computing the electronic structure with both LDA and PBE. In Fig. 4.10 the computed electronic band structure is depicted together with FP-LAPW/PBE data. It can be seen that our computed band gap of 0 eV corresponds to other PBE data while the experimental band gap is 0.7 eV [165].

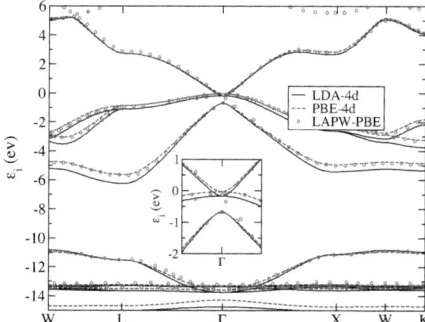

Figure 4.10: Electronic band structure of InN computed with LDA and PBE. The 4d semicore states are treated as valence electrons. The top of the valence band has been aligned to the Fermi level (0 eV). The red circles show LAPW-PBE data taken from Ref. [166] which are in good agreement with our pseudo potential results. The inset magnifies the area around the direct band gap at the Γ point.

Discussion

In this section we presented the computed phonon dispersion curves of the investigated III-V semiconductor systems and laid out the foundation for computing their thermodynamic properties. For both exchange-correlation functionals LDA and PBE our obtained acoustic phonon branches are in good agreement with experimental results as well as other first-principles DFPT-LDA studies. Since in the temperature regime we are focusing on in this work the acoustic branches are dominant when computing the free energy surface, our phonon spectra should be a reliable basis for deriving the thermodynamic properties. Also the optical branches obtained from our first-principles calculations are qualitatively in good agreement with experimental and other theoretical data.

With a relatively small super cell (2x2x2, fcc cell, 64 atoms) it is possible to reproduce the available phonon spectra obtained from DFPT-LDA calculations with the direct approach well with deviations of less than 2 meV. The achieved accuracy of the forces[4] allows to determine the phonon dispersion curves for all investigated systems with only small deviations to the experiment (typically below 3 meV). For some systems, however, there are discrepancies between our data and the experimental ones. For example, for AlAs the TA branch at X is 4 meV above our data, the frequencies at the L point of GaP is underestimated by about 4 meV compared to the neutron scattering data. These deviations are obtained consistently with LDA and PBE. Our data are, however, qualitatively and quantitatively very close to the DFPT-LDA frequencies at these points as well as in their vicinities.

There is also a small deviation in the predicted behavior at the vicinity of Γ between our TA branches on those obtained with DFPT-LDA. Typically, our branches are slightly steeper than the DFPT-LDA ones. This effect is, however, small (between Γ and K the deviation is always less than 3 meV) and the additional computational effort of using a 3x3x3-fcc super cell cannot be justified.

Our approach has *minor* difficulties in the optical branches, though. In our ansatz we neglect the coupling between phonons and electric fields which is responsible for the degeneracy at the Γ point. This effect,

---
[4]All forces have been converged to a *numerical* accuracy of $\Delta F_{x,y,z} \leq 1e^{-6}$ H/Bohr.

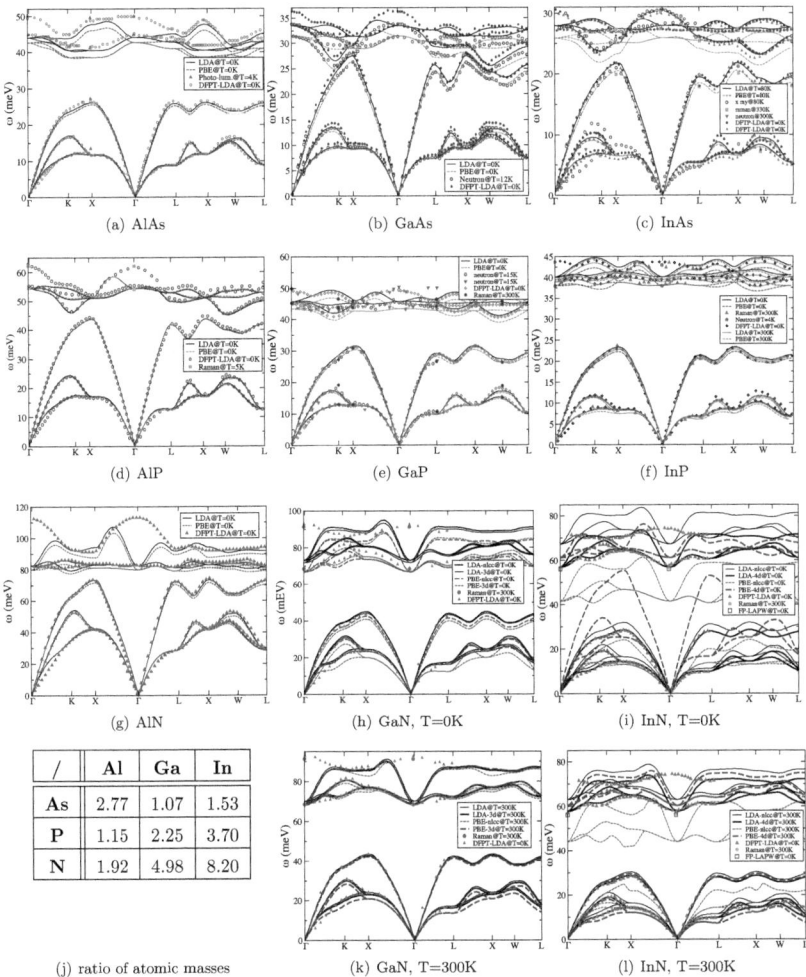

Figure 4.11: (a)-(i), (k)-(l) Influence of the atomic mass ratio to the phonon dispersion curves. Large mass ratio (e.g., InP, GaN, InN) lead to large phonon band gaps and flatter optical branches while smaller ratios (e.g. GaAs, AlP, InAs) lead to small gaps and larger dispersions of the optical branches.

however, can be neglected here since only a very small region in **q**-space is affected and the red-shifted LO branches are energetically very high (typically above 80 meV) and will be only thermodynamically excited close to the melting point. Our data are able to reproduce the typical trends [47, 167] of the phonon spectra: With increasing mass *differences* the phonon band gap increases and the optical phonons become flatter (see Fig. 4.11). Considering a linear di-atomic chain (Fig. 4.12) with masses M and m the gap between acoustic and optical branches is driven by the mass ratio. With larger gaps the optical phonons become flatter [47].

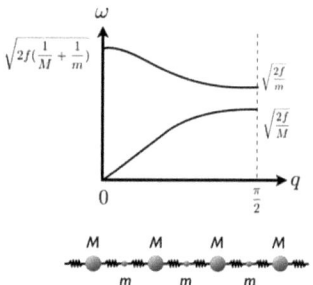

Figure 4.12: Phonon spectra of an infinite di-atomic linear chain. The chain is formed with atoms of masses M and m. If $M > m$ a gap opens at $q = \frac{\pi}{2}$. Larger mass differences open the gap more while the optical phonon branch becomes flatter along the path $0 - \frac{\pi}{2}$ [47].

For systems with relatively small mass ratios (e.g., GaAs ($\frac{M}{m} = 1.07$), AlP ($\frac{M}{m} = 1.15$), InAs ($\frac{M}{m} = 1.53$) the TO mode shows a oscillatory behavior while for systems with larger mass ratios (e.g., GaP ($\frac{M}{m} = 2.25$), AlAs ($\frac{M}{m} = 2.77$), InP ($\frac{M}{m} = 3.70$), GaN ($\frac{M}{m} = 4.98$), InN ($\frac{M}{m} = 8.20$) the TO branch becomes flatter. The obtained frequency range of the optical phonons increases in the order (1) arsenides, (2) phosphides, and (3) nitrides. Within these material classes the frequencies increase with the choice of the cation in the order In, Ga, and Al.

For all investigated systems except GaN and InN we obtained phonon dispersions with LDA and PBE which are very close to each other. We have shown that the applied method is capable of computing accurately phonon spectra for a wide range of systems if the pseudo potentials are constructed carefully. For GaN and InN, the choice of the treatment of the d-semicore states (either via NLCC or by explicit treatment as valence) is crucial. In particular, a good description of the band gap is necessary to obtain accurate phonon spectra which can be achieved with LDA and PBE-NLCC. InN, which is known to be challenging in particular for PBE, the phonon spectra deviate significantly which might lead to a substantial error bar in the computed thermodynamic properties. Following the previous discussion (p. 122) an application of more advanced (hybrid) functionals would be interesting but would exceed the scope of this work.

4.3.3 Thermal expansion

The exact knowledge of the thermal expansion behavior of semiconductors is technologically very important. For example, when growing semiconductors on substrates it is crucial to consider the thermal lattice mismatch between the substrate and the semiconductor. The same applies for semiconductor interfaces. Therefore, in this section the thermal expansion for the 9 investigated semiconductor systems will be studied. With the volume derivatives of the phonon spectra $d\omega_i(\mathbf{q})/dV$ the thermal expansion coefficients $\alpha(T)$ can be obtained via Eqs. (2.67), (2.70), (2.71), and (2.80). Since all phonon dispersion curves have been computed with LDA and PBE it is possible to estimate their performances when computing linear expansion coefficients of III-V semiconductors. In Fig. 4.13 we present the obtained expansion coefficients in the temperature range between 0 K and the melting temperature of the corresponding systems.

In Fig. 4.13(a) the computed linear expansion coefficients (Eq. (2.80)) of **AlAs** obtained with LDA and PBE are compared with experimental data [168, 169] as well as another pseudo potential plane-wave LDA simula-

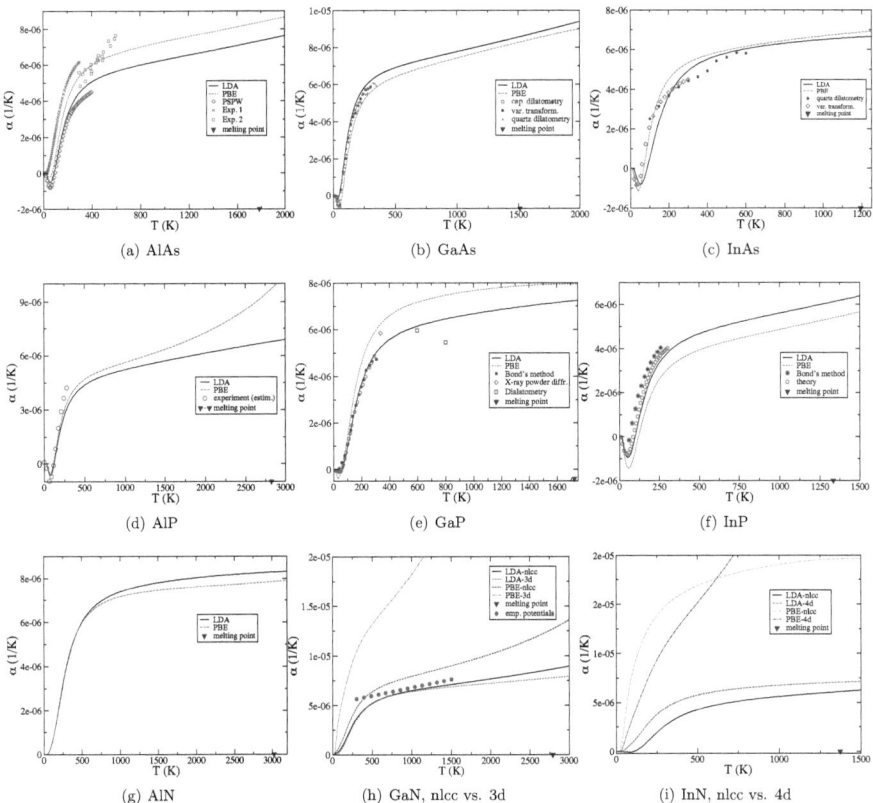

Figure 4.13: Temperature dependence of the linear expansion coefficients $\alpha(T)$ of all investigated III-V semiconductors. We compare results obtained from LDA (solid black lines) with PBE (red dashed lines). The accuracy of $\alpha(T)$ in the high temperature limit strongly depends on the quality of the description of the minimum of $\alpha(T)$. Minor error bars in the location or amplitude of the minima leads to significant shifts of $\alpha(T)$.

tion [170]. There is a substantial scattering of the experimental data sets indicating significant uncertainties for this material. Furthermore, only data between 0 K and 400 K are available. Our LDA and PBE curves show virtually identical slopes. The shift of $1 \cdot 10^{-6} K^{-1}$ in the high temperature limit can be explained with the slightly differently pronounced minima of $\alpha(T)$. The location of this minima at 55 K (LDA) and 50 K (PBE) is in very good agreement with the LDA calculation from Ref. [170] (52 K).

Similarly good results have been obtained for **GaAs**. As for AlAs the slopes of $\alpha_{\text{LDA}}(T)$ and $\alpha_{\text{PBE}}(T)$ are very close to each other. In case of GaAs the description of the location of the minimum as well as its amplitude is also in good agreement ($\Delta T < 8\,K$). The experimental values, obtained with capacitance dilatometry [171], variable transformer measurements [172], and quartz dilatometry [173], are always well in between $\alpha_{\text{LDA}}(T)$ and $\alpha_{\text{PBE}}(T)$.

125

The scattering of the experimental results (quartz dilatometry [174], variable transformer measurements [172]) for **InAs** reveals uncertainties for this system. Our data predict the location of the minimum with a small deviation of 9 K between LDA and PBE and agrees very well with the experimental dataset. At higher temperatures the curvatures of α_{LDA} and α_{PBE} differ slightly. The scattering between the experimental datasets make a verification of our data difficult.

The second row of Fig. 4.13 depicts the temperature dependencies of the thermal expansion coefficients of the investigated phosphides. For **AlP** the graph labeled as experiment is only a "rough estimation" (see [175]) based on an extrapolation of other III-V semiconductor zincblende structures. Its quality is questionable and can only serve as a check of the order of magnitude. The AlP phonon spectra for LDA and PBE are in very good agreement which is reflected in the high accuracy of the prediction of the minima at $T_c = 72\,\text{K}/78\,\text{K}$ (LDA/PBE). AlP has a very high melting point at T=2823 K. At such high temperatures unharmonicities beyond the *quasi*-harmonic approximation are expected to play an important role [37]. The high temperature results should, therefore, only be taken as crude approximation.

In contrast to most of the other tetrahedrally bounded III-V semiconductors investigated here, previous studies predicted that **GaP** does not exhibit an anomaly in the thermal expansion. Soma *et al.* [41] reported first that the magnitude of the TA mode Grüneisen parameters in GaP are too small to generate the anomaly. Deus and co-workers [42] confirmed Soma's theoretical prediction using Bond's method [176]) with a relative error of $2 \cdot 10^{-5}$. However, in 1986 Haruna *et al.* [43] were able to measure a tiny anomaly effect in GaP at low temperatures with a relative error of only $2 \cdot 10^{-6}$. In Fig. 4.13(e) it can be seen that our results also show that there is a shallow minimum of the linear expansion coefficients at $T_c = 32\,\text{K}/42\,\text{K}$ (LDA/PBE). The experimental value of $T_c = 38\,\text{K}$ lies between our results for T_c^{LDA} and T_c^{PBE}.

InP and GaP show a good agreement with respect to the slope and the location of the minimum. It is noteworthy that our LDA data are in both cases slightly closer to the experimental values. Our data agree well with results obtained by Bond's method [40] as well as a DFPT-LDA result [170].

The last row of Fig. 4.13 presents the nidrides. **AlN** has the highest melting point of all considered systems studied in this work. It melts at 3025 K. Our LDA and PBE curves in Fig. 4.13(g) are virtually the same between 0 K and 750 K. Considering the good agreement of the corresponding acoustic phonon dispersion curves this is not surprising. At high temperatures (above 1000K) the PBE results have a slightly smaller slope than the LDA expansion coefficient curve. The deviation at the melting point is less than $0.5 \cdot 10^{-6}\,K^{-1}$ and thus well within acceptable limits.

The example of the temperature dependence of the thermal expansion coefficients of **GaN** shows nicely how the quality of the computed phonon spectra influences the accuracy of $\alpha(T)$ and provides an estimation of the limits of the applied method. In Fig. 4.14a) and b) the phonon spectra which have been computed earlier as well as the linear expansion curves are presented. As discussed above in case of LDA the influence of the choice of the d-semicore treatment (NLCC or with 3d valence electrons) to the phonon spectra is small. Both $\omega_{\text{LDA}}^{\text{nlcc}}(\mathbf{q})$ and $\omega_{\text{LDA}}^{3d}(\mathbf{q})$ are very close to one another and agree very well with other DFPT-LDA data. From Fig. 4.14(b) it can be seen that both curves $\alpha_{\text{LDA}}^{\text{nlcc}}(T)$ and $\alpha_{\text{LDA}}^{3d}(T)$ are almost identical. Also the phonon dispersion curves obtained with PBE-NLCC are in good agreement, even though they are slightly red-shifted. The corresponding linear expansion coefficients $\alpha_{\text{PBE}}^{\text{nlcc}}(T)$ respond to this red shift with a larger slope in the temperature interval between 0 K and room temperature. Above that the slope of $\alpha_{\text{LDA}}(T)$ and $\alpha_{\text{PBE}}^{\text{nlcc}}(T)$ are the same with a constant shift of $1.8 \cdot 10^{-6} K^{-1}$. As discussed already above, by improving the description of the d-semicore states (treating them in the valence) the prediction of the electronic band gap of GaN gets worse which leads to a significant red shift of the phonon frequencies ω_{PBE}^{3d}.

The last investigated system, **InN**, is known to be challenging for DFT. Above the obtained results for the phonon spectra for LDA and PBE have been presented. We discussed that PBE has significant shortcomings when describing InN, in particular, we focused on the electronic band gap and the phonon dispersion curves. The latter ones are crucial for computing the free energy $F(V,T)$ and thus, the linear expansion coefficients $\alpha(T)$. In Fig. 4.15(b) we present our data for $\alpha(T)$. It can be seen that LDA-NLCC and LDA-4d yield almost the same slope, only in the low temperature limit the description of the d-semicore states influences the location of the anomaly of the temperature dependence of α, with LDA-NLCC we obtain $T_c = 78\,K$ while an explicit treatment of the semicore states in the valence yields $T_c = 26\,K$. The deviation between $\alpha_{\text{LDA}}^{\text{nlcc}}(T)$ and $\alpha_{\text{LDA}}^{\text{4d}}(T)$ at the melting point $T_m = 1373\,K$ is less than $0.9 \cdot 10^{-6} K^{-1}$ and thus, still within acceptable limits.

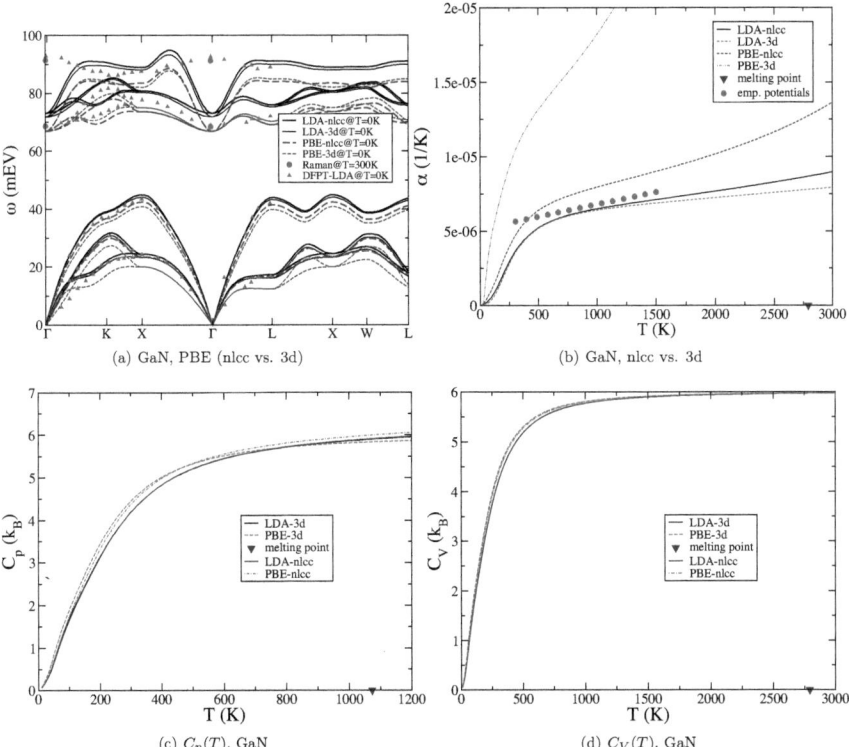

(a) GaN, PBE (nlcc vs. 3d)

(b) GaN, nlcc vs. 3d

(c) $C_p(T)$, GaN

(d) $C_V(T)$, GaN

Figure 4.14: Influence of the explicit treatment of the d-electrons to all investigated properties of GaN. (a) The obtained phonon spectra of GaN provide an indication about the quality of the description of the linear expansion coefficients. A poor description of the phonon frequencies suggests also significant problems for evaluating derived entities, such as $\alpha(T)$. (b) Temperature dependence of the linear expansion coefficients $\alpha(T)$ of GaN computed with LDA and PBE. The d-semicore states have been treated in the core using NLCC as well as explicitly in the valence (3d). (c) and (d) show the heat capacities $C_p(T)$ and $C_V(T)$, respectively.

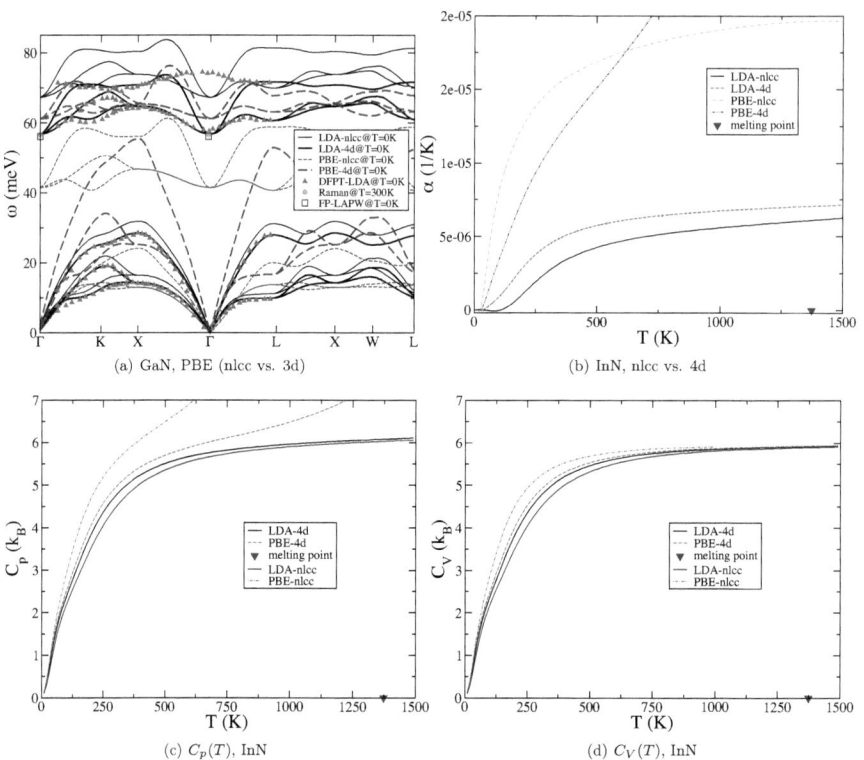

Figure 4.15: Influence of the explicit treatment of the d-electrons to all investigated properties of InN. (a) The phonon spectra of InN as well as the (b) linear expansion coefficients versus temperature. (c) and (d) show the heat capacities $C_\mathrm{p}(T)$ and $C_\mathrm{V}(T)$, respectively. LDA (NLCC and 4d) is able to describe the phonon band structure of InN better which results on a reasonable description of the thermal expansion behavior of cubic InN. The huge errors introduced by PBE (both NLCC and 4d) influences the qualitative picture of the thermal expansion drastically.

Discussion

All investigated systems exhibit a thermal expansion anomaly in the low temperature regime, typically between 20 K and 80 K. In this low temperature interval most of the systems show negative thermal expansion coefficients (see explanation above, Sec. 4.2.1). An accurate description of the thermal expansion coefficients requires very accurate values of $d\omega_i/dV$. The accuracy of the temperature slope as well as the location of the minimum is mainly determined by the acoustic phonons (TA, TA1, TA2).

The numerically very sensitive minimum determines whether $\alpha_\mathrm{LDA}(T)$ or $\alpha_\mathrm{PBE}(T)$ is the upper or lower limit. For example, in case of AlAs $\alpha_\mathrm{LDA}(T) < \alpha_\mathrm{PBE}(T)$ while for GaAs $\alpha_\mathrm{LDA}(T) > \alpha_\mathrm{PBE}(T)$. The thermal anomaly makes it, therefore, difficult to derive a trend whether LDA under- or overestimates $\alpha(T)$ of zincblende III-V semiconductors.

The expansion coefficient curves of GaN and InN which have been obtained with PBE-NLCC or explicit treatment of the 4d electrons show a qualitatively wrong dependence. This is likely due to the description of the band gap (see p. 119). The effect on the phonon spectra (and their volume derivatives) induce difficulties in the description of the linear expansion coefficients of GaN and InN.

4.3.4 Heat capacity

Besides the thermal expansion coefficients this work focuses also on a derivation of the temperature dependence of the heat capacity $C_p(T)$ and $C_V(T)$ from first-principles. In case of $C_p(T)$ the theoretical results can be easily compared with the experimental data, since $C_p(T)$ can be measured directly.

The general shape of the temperature dependence of the heat capacity is determined by two limits. According to the Debye model at low temperatures $C_V(T)$ should scale as T^3 while in the high temperature region $C_V(T)$ converges to the Petit-Dulong limit $6k_B$. In between those limits $C_V(T)$ is determined by the details of the atomic vibrations.

In Figs. 4.16, 4.17, and 4.18 the heat capacities of the arsenides, phosphides, and nitrides are presented, respectively. For zincblende-**AlAs**, **-AlP**, and **-AlN** no experimental data were found which makes a verification of our results difficult.

In Fig. 4.16c-d we compare our computed $C_p(T)$ as well as $C_V(T)$ with both experimental data and other theoretical data obtained by first-principles methods employing LDA pseudo potentials for **GaAs**. In the temperature regime between 0 K and 500 K our results show the same behavior as the experimental data (taken from Ref. [177], experiments are labeled as in this reference). Due to the large scattering between the performed experiments a conclusive evaluation of our data is not straightforward. Thus, we compare our data with other LDA pseudo potential data [178] which are virtually identical to our data.

The picture for **InAs** is reminiscent to GaAs. The various experimental data sources (taken from Ref. [177], experiments are labeled as in this reference) differ in their results at high temperatures above 400 K. At higher temperatures the theoretical $C_p(T)$ curve remains more flat then the experimental ones. Furthermore, the agreement between LDA and PBE is obtained only up to 600 K.

In Fig. 4.17 we present the theoretically obtained temperature dependencies of the heat capacity at constant pressure and constant volume for AlP, GaP, and InP, respectively. For all investigated **phosphides** the agreement between LDA and PBE is very good, indicating that the error bar introduced by the XC functional is very small for the heat capacity.

Discussion

A direct comparison of our result with experiment is difficult due to large scattering between various experimental data sets. In Ref. [177] a comprehensive overview on the different challenges in performing accurate experiments focusing on measuring the heat capacities for GaAs and InAs is presented. Various sources of error are responsible for a significant discrepancy among various experiments of InAs and GaAs, such as the choice of the calorimeter container material[5], introduction of additional heat when entering the test ampulle into the container, accounting for thermal dissociation effect. Such issues introduce large uncertainties in the experimental measurements and must be considered when performing the comparison to theory.

---
[5]The $C_p(T)$ curve for GaAs labeled "4: drop calometry" differs significantly from the other graphs probably due to the choice of Ta as container material. Ta reacts with both GaAs and InAs.

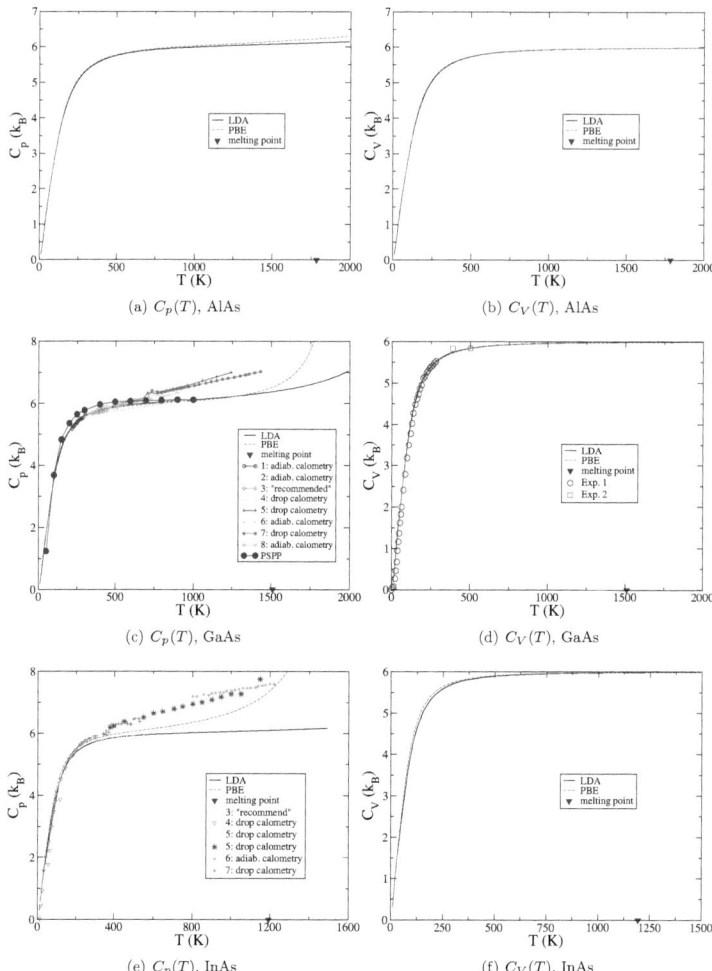

Figure 4.16: $C_p(T)$ and $C_V(T)$ for AlAs, GaAs, and InAs.

Also the verification of our data with measurements of the phosphides is not straightforward since measuring the heat capacities of phosphides is challenging. Phosphorus exhibits an allotropic behavior, i.e., it can exist in different forms. A famous example of allotropy is carbon which can exist in the diamond and graphite form. For phosphorus 9 allotropes are known at this time [181, 182, 183, 184, 185]. Since they have different enthalpies of formation, a misinterpreted form of P leads inevitably to wrong data. These experimental challenges are responsible for the scattered data in Fig. 4.17.

Qualitatively for all investigated systems except GaN and InN we obtained similar data for $C_p(T)$ and

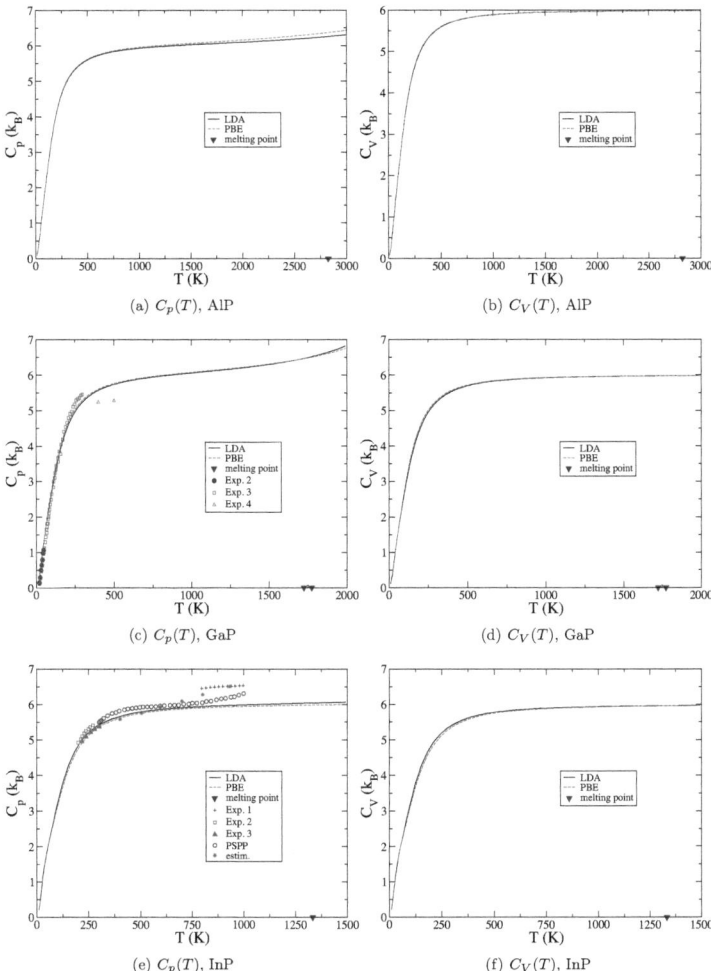

Figure 4.17: $C_p(T)$ and $C_V(T)$ for AlP, GaP, and InP. AlP that is meta-stable in the zincblende phase no experimental data were found. The experimental calometry data for GaP have been taken from Ref. [168]. In case of InP the calometric data were taken from Ref. [179]. The InP DFT-LDA pseudo potential data labeled "PSPP" are taken from Ref. [180].

$C_V(T)$ with both XC functionals. The phonon and electronic band structure of cubic GaN and InN can be only accurately described within LDA. PBE introduces errors in the band gap as well as a significant red shift of the phonon frequencies. This inaccuracy enters the free energy $F(T,V)$ and hence, the derived thermodynamic properties such as $\alpha(T)$ as well as $C_p(T)$ and $C_V(T)$. In Figs. 4.14 and 4.15 all these entities are depicted together. The temperature dependencies $C_p^{\text{LDA,nlcc}}(T)$ and $C_V^{\text{LDA,3d}}(T)$ show qualitatively the

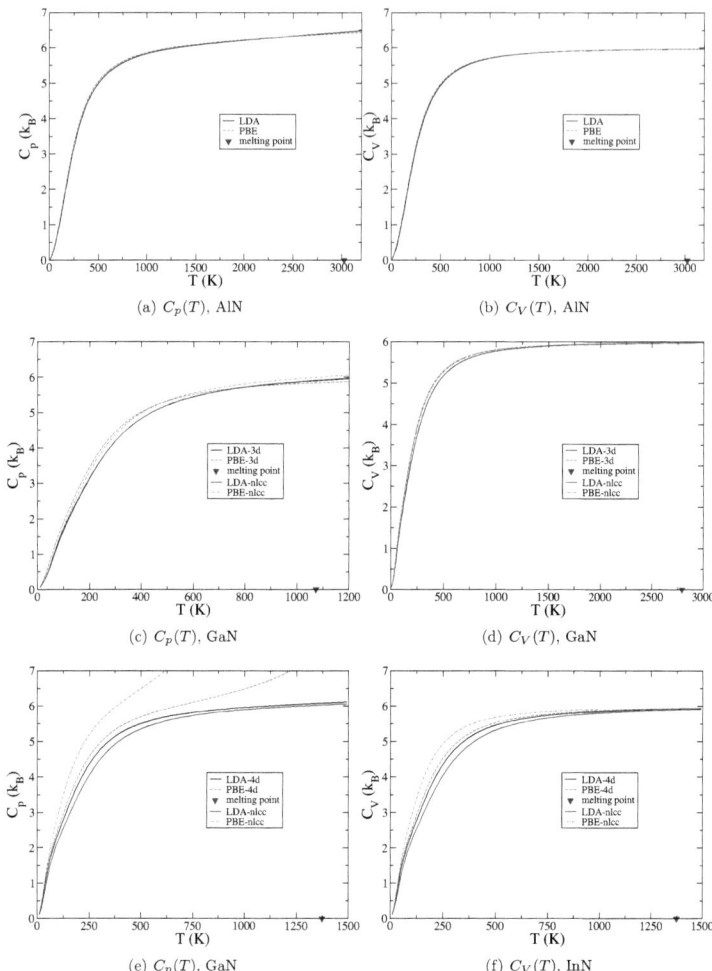

Figure 4.18: $C_p(T)$ and $C_V(T)$ for AlN, GaN, and InN. All systems are described well within LDA while PBE shows problems in describing the heat capacities of GaN and InN.

same behavior which is due to the good agreement of $\omega_{\text{LDA}}^{\text{nlcc}}(\mathbf{q})$ and $\omega_{\text{LDA}}^{\text{3d}}(\mathbf{q})$.

4.3.5 Conclusions

In this chapter the results of the computations of thermodynamic properties have been presented. All necessary calculations have been performed with the S/PHI/nX package in order to demonstrate that the abstract and complex S/PHI/nX approach is capable of combining a great degree of development flexibility with high performance calculations of realistic systems. As benchmark we have chosen the computation of thermodynamic properties of III-V semiconductors.

We found that for all investigated systems the application of LDA provides a good basis to obtain high accuracy phonon spectra $\omega_i(\mathbf{q})$, linear expansion coefficients $\alpha(T)$ as well as heat capacities $C_{p,V}(T)$. In case of the two systems GaN and InN, PBE introduces major difficulties when computing thermodynamic properties. We have demonstrated that even minor deviations in the phonon frequencies introduce problems in the prediction of derived entities such as $\alpha(T)$ and $C_{p,V}(T)$. Therefore, we have shown that it is crucial to perform very thorough convergence tests with respect to the pseudo potential, the energy cut-off and the **k**-point sampling with respect to the phonon dispersion curves. By means of GaAs we demonstrated that deriving thermodynamic properties requires dramatically higher convergence criteria than, e.g., when computing electronic structures.

The obtained phonon spectra follow the general trends: Larger ratios of atomic masses tend to large phonon gaps and flatten the optical phonon dispersions. The III-V semiconductors in the zincblende phase have negative mode-Grüneisen parameters $\gamma(\mathbf{q})$ which are occupied at low temperatures (typically at $T < 100\,\mathrm{K}$). In this regime they show an anomaly in the thermal expansion. At increasing temperatures the crystal compresses. Above the critical temperature positive mode-Grüneisen parameters dominate and normal expansion behavior occurs. A theoretical description of this anomaly based on first-principles is extremely sensitive since it can introduce shifts of $\alpha(T)$ in the high temperature regime. For all systems it was possible to keep this shift within acceptable limits ($\Delta\alpha < 2 \cdot 10^{-6}\,\mathrm{K}^{-1}$) indicating a very high numerical accuracy of our data.

Chapter 5

Conclusions and Outlook

In this work we derived and implemented a new physics meta-language to develop highly efficient programs in the field of computational materials design (CMD). It simplifies the development process of *ab-initio* based multiscale approaches drastically. Our meta-language provides intuitive language elements to express algebraic equations, quantum mechanical expressions in the Dirac notation, and an efficient representation of equations of motions. *State-of-the-art* programming techniques from the field of computer science have been developed / derived in this work in order to automatically create optimized machine code.

Our meta-language supports the application of Dirac's notation. Therefore, the "building blocks" of the Dirac notation, i.e., Dirac vectors, projectors, and operators have been introduced as generic data types (p. 80). Our concept considers, in particular, future extensions with respect to the implementation of modern basis-sets and Hamiltonians. Therefore, we derived a technique to support virtual template projector functions (p. 81). It allows the compiler to determine the quantum mechanical context and thus, to replace complex expression with highly optimized function calls during the compilation. Benchmarks that compare the run-time performance of various representative calculations with VASP demonstrate that with our solution an intuitive Dirac-notation interface can be combined with high executional performance (p. 104).

With the new meta-language complex algebraic or quantum mechanical algorithms can be developed easily with only rudimental programming skills. However, the high abstraction level requires a smart environment to identify typical problems during the development process. Therefore, we developed an automatic error detection mechanism which is capable of identifying typical program inconsistencies in quantum mechanical or numerical algorithms (see Sec. 3.1.4). With this mechanism quantum mechanical expressions which are unphysical but syntactically correct can be identified instantly which decreases the time necessary for the development process drastically.

In order to compute the material properties related to the electronic structure of a system, an efficient library specialized in electronic minimization to obtain the Born-Oppenheimer surface has been developed (pp. 41). In the current version of our library we introduced an exponentially converging all-band preconditioned conjugate-gradient algorithm for semiconducting and insulating systems (Sec. 41). We also introduced an state-by-state preconditioned conjugate-gradient minimizer which employs DIIS charge density mixing (pp. 43). This scheme can be applied to systems with empty or partially occupied states, such as metals.

The computation of structural material properties requires an efficient representation of equations of motion. Therefore, we derived transformation pipelines (Sec. 3.3.2) which allow a separation of the multi-dimensional minimization schemes from structural constraints or frequency filters. Due to our automatic BLAS/LAPACK

CHAPTER 5. CONCLUSIONS AND OUTLOOK

mapping (p. 68) the resulting library to represent atomic structures performs also efficiently for large atomic systems described by (semi-)empirical potentials. It provides the major representations of atomic structures, namely the *xyz-* and the *degree of freedom*-form. In our approach both forms can be applied *simultaneously*.

Codes developed in the S/PHI/nX meta-language can be written very transparently and remarkable short which simplifies the process of code development significantly. For example, the entire DFT Hamiltonian in S/PHI/nX requires only 550 code lines.

In order to guarantee computational peak performance we developed a highly efficient numeric library SxMath which is the foundation of the S/PHI/nX project. It provides a *functional* interface reminiscent to high-level toolkits like Mathematica. In order to combine an intuitive interface with high executional speed, performance problems that occur typically in functional approaches (*abstraction penalties*, Sec. 3.1.3) had to be addressed. Therefore, we derived new techniques such as S/PHI/nX type mappers (p. 61). They ensure computationally optimal[1] data types to allow effective computations on temporary algebraic objects. With a reference counting technique specialized for algebraic vectors/matrices (p. 63) we are able to replace procedural with the more transparent functional interface without any performance loss. The often tedious and error-prone task in High-Performance computing (HPC) of BLAS/LAPACK function call mapping can be accomplished fully automatically due to generic programming techniques using template classes (p. 60). With this technique *all* algebraic expressions are mapped consequently to the available HPC function calls. Our algebra library "SxMath" is applicable for a wide range of applications which rely on highly efficient evaluation of algebraic algorithms. It could be shown that our approach improves the run-time performance of typical algebraic expressions necessary in DFT program packages dramatically (p. 68) compared to standard numerical toolkits (e.g. Blitz++, Boost).

In the current version of S/PHI/nX we focused, in particular, on heavy usage of blocking algorithms, i.e., the formulation of the computationally demanding algorithms in a matrix-matrix form. In this notation matrices can be consequently subdivided such that they fit into the steadily increasing level-caches of modern computer architectures, which guarantees peak-performance. For the group of investigated III-V semiconductors we conducted benchmarks with the widely applied VASP package. Even for systems which require larger (converged) energy cut-offs, where the pseudo-potential plane-wave approach is computationally more demanding than PAW, the high optimization level of S/PHI/nX is comparable and for some systems even faster than VASP.

By organizing S/PHI/nX as library instead of a single monolithic program package, code fragments can easily be reused. S/PHI/nX add-ons (Sec. 3.3.3) are small programs (usually less than 50 or 150 lines) which can be developed in very short times. An add-on has full access to all elements of the above described class hierarchy. In order to analyze results wave functions can be imported, projected on other basis-sets, densities of state can be obtained, or complex atomic structures can be generated. For the most common tasks for the preparation as well as the analysis a set of 50 S/PHI/nX add-ons have been developed. The add-ons have a common interface so that the output of one add-on can act as input of another. Complex analysis pipe lines can be easily created by the user. It is worth mentioning that S/PHI/nX has been developed while strictly obeying various standards, such as ANSI, POSIX, as well as a source code style guide applied in industrial environments. The code is therefore platform independent and is available on Linux, MacOS X, FreeBSD, AIX, HPUX, and Windows. The S/PHI/nX testbed ensures reliability during the development phase. Code consistency is accomplished using a reviewing process.

In the second part of this work it was tested whether the new S/PHI/nX package is sufficiently fast and

---

[1]optimal = smallest required accuracy and byte width

accurate to be applied for realistic systems by means of thermodynamic properties of III-V semiconductors. We could confirm the findings of Ref. [37] that the computation of thermodynamic properties from first-principles requires high convergence in all parameters well beyond what is commonly needed to T=0 K properties. In this study calculations within LDA and PBE have been conducted in order to study the influence of the exchange-correlation potential to the accuracy of the obtained thermodynamic properties. We investigated the temperature dependencies of the linear expansion coefficients $\alpha(T)$ as well as the heat capacities $C_p(T)$ and $C_V(T)$ in the temperature regime between 0 K and the corresponding melting temperatures. For all investigated systems we found a good agreement with experimental data and we could establish that the direct approach is a reliable method to obtain thermodynamic properties from first-principles. All obtained phonon spectra are an good agreement with the experiment ($\Delta\omega_i(\mathbf{q}) \leq 4\,\mathrm{meV}$). Generally, LDA and PBE provide almost identical phonon dispersion curves. In our study we obtained phonon frequencies of the acoustical phonons very close or virtually identical to experiments or other theoretical investigations. The very sensitive temperature dependence of the linear expansion coefficients $\alpha(T)$ have been reproduced well. The magnitude of the thermal expansion anomaly of the cubic III-V semiconductors could be well reproduced within LDA. We found that PBE introduces minor deviations in the linear expansion coefficients and heat capacities of GaN and InN. This is likely related to the description of the electronic structures of both structures. A further investigation with more advanced exchange-correlation functionals such as hybrid functionals would be interesting. At high temperatures close to the melting point our computed $\alpha(T)$ and $C_{p,V}(T)$ results deviate from the experimental data. Therefore, an investigation which also includes unharmonic effects would clarify whether the quasi-harmonic approximation fails for these systems in the high-temperature regime.

The performed calculations make the upcoming next steps clear. The application of norm-conserving pseudo potentials is computationally too demanding and the treatment of the d semicore states of Ga and In introduces problems. Therefore, the implementation of PAW as new basis-set is imperative for the S/PHI/nX project. Furthermore, our benchmarks have shown that the parallelization of S/PHI/nX is urgently needed to perform simulations of even larger simulation cells. Due to the required high energy cutoffs in this work we had to restrict ourselves to a 64 atom cell which introduced a *minor* phonon softening. For other systems, however, the influence might be more important and the simulation of larger cells might be necessary. Besides improved basis-sets also parallelization of S/PHI/nX is, therefore, an urgent item on the list of features that will be implemented in the upcoming version(s). We believe that the *template linkage* technique can be extended such that a semi- or even fully automatic MPI parallelization becomes possible. Similar to the automatic BLAS/LAPACK mapping or the Dirac-notation we think that this technique can also be used to map algebraic or quantum mechanical expressions to proper MPI calls during the compilation. The distribution of data, which determines the performance and scalability of the parallel approach, depends on the basis-set. Since that is already known to our *template linkage* approach, a high performance and automatic MPI parallelization might be possible. First proofs of concepts have indicated that such an approach can be successful.

With S/PHI/nX we introduce a new flexible development framework and a user-friendly program package for CMD, that has been successfully applied already to a variety of investigations. S/PHI/nX has been used to

- compute bio-inspired systems systems such as polyalanine alpha and π−helix [186, 187, 188, 189], crystalline α-chitin [190]

- computation of thermodynamic properties of metallic systems, e.g., of Al [191] and Fe [192] up to the

CHAPTER 5. CONCLUSIONS AND OUTLOOK

melting points

- compute electro-optical properties of quantum dots [193, 194, 195] and quantum wells [196, 197, 198],

- investigate material properties of semiconductors, e.g., dislocations in wurtzite GaN [199], application of maximally-localized Wannier functions to III-V semiconductors [200] or to semiconductor alloys [201], description of nitrogen solubility at GaAs and InAs (001) surfaces [202, 203]), compute finite-size corrections for charged defect supercell calculations [204], investigate ferromagnetic systems such as GaMnAs [205]

- address the band gap problem of DFT using EXX [165, 206, 207, 208, 209, 210, 211, 212, 213]

- introduce an efficient all-band conjugate gradient method for metallic systems [214] and a plane-wave implementation of the real space k·p formalism and continuum elasticity theory [215]

- investigate the role of anharmonic effects on the elasticity of ice [216]

- perform atomic-scale spin-polarized scanning microscopy simulations of nonmagnetic metallic surfaces [60]

Besides the actual program, the S/PHI/nX C++ library has been used separately to develop new tools:

- In Ref. [217, 218] the S/PHI/nX library has been applied to implement tools for the ABINIT project to simulate STM and STS simulations on magnetic and non-magnetic metallic surfaces.

- Based on the efficient numeric libraries of S/PHI/nX a powerful graphic render engine for interactive scientific visualization could be introduced [219].

- The network communication and file i/o libraries have been used to create an efficient generalized database for multi-physics applications [220].

The S/PHI/nX project is also engaged in simplifying the exchange of data between research groups within the Ψ_k-network community. For example, in S/PHI/nX a general file format has been introduced which allows a comfortable exchange of huge data sets such as wave functions and potentials. Based on the S/PHI/nX format the Ψ_k-network introduced the ETSF_IO [221] data exchange format. With this format a standard for exchanging wave functions and potentials between various plane-wave codes has been established[2].

---

[2]S/PHI/nX supports currently both file formats, the original S/PHI/nX format "sxb" as well as ETSF_IO.

Acknowledgments

A complex project such as S/PHI/nX can only be realized with the support and collaboration of many people.

I'm grateful to my supervisor Jörg Neugebauer for giving me the opportunity to work on this exciting project. With patience he introduced me to the field of DFT and method development. He provided perfect working conditions and created such a nice atmosphere in his group that I could develop S/PHI/nX with great joy.

I would like to thank the S/PHI/nX team for all their contributions: Hazem Abu-Farsakh, Abdullah Al-Sharif, Alexey Dick, Christoph Freysold, Lars Ismer, Abduallah Qteish, and Matthias Wahn. Here, I thank, in particular, Christoph Freysold for all those daily discussions and his valuable contributions across the package and Alexey Dick who spent an immense time in the unpleasant task of bug tracking and his vast contributions to the stability of the code.

I would like to extend my thanks also the new members of the S/PHI/nX team Vaclav Bubnik, Björn Lange, Oliver Marquadt, Gernot Pfanner, and Thomas Uchdorf. With the on-going projects they show that the project continues to prosper.

I am grateful to Lutz Schützenmeister who woke my interest in physics in the first place. The S/PHI/nX package wouldn't have none of its beauty without Roland Augst who introduced me to the fundamental concepts of code design.

For proof reading and valuable discussions I would like to thank Alexey Dick and Tilmann Hickel.

I would like to thank my fiancée for her constant support in all the years of the S/PHI/nX development and the never-ending time of writing up. I want to thank my parents for waking my interest for natural and computer sciences. Without their constant effort in motivating me I might not have been able to complete this project.

Appendix A

Computational details

A.1 Pseudo potentials

The pseudo potential have been generated with the following configurations and cut-off radii[1] (r_s^{cut}, r_p^{cut}, r_d^{cut}, r_f^{cut})

Al: $3s^2 3p^1 3d^0$ (1.4, 1.5, 1.4)

Ga (nlcc): $4s^2 4p^1 4f^0$ (2.3, 2.4, 2.2, 2.8)

Ga (3d): $3d^{10} 4s^2 4p^1 4f^0$ (2.3, 2.4, 2.2, 2.8)

In (nlcc): $5s^2 5p^1 5g^0$ (2.3, 2.4, 2.2)

In (4d): $4d^{10} 5s^2 5p^1 5g^0$ (2.3, 2.4, 2.2, 2.8)

P: $3s^2 3p^1 3d^0$ (1.8, 2,2, 1.9)

As: $4s^2 4p^3 3d^0$ (1.4,1.2, 2.1)

N: $2s^2 2p^3 3d^0$ (1.5, 1.5, 1.5)

In case of NLCC we used partial core densities with cutoff radii 1.3 (Al, Ga), 1.8 (In)

The reference energies for the unbound unoccupied states were set to $2p$ (N), $3p$(Al) eigenvalues, and to 5 eV (Ga 4d), 0 eV (Ga 4f), 15 eV (In 5d), -30 eV (In 5g)

A.2 Convergence parameters

	E_{cut} (Ry) Al	E_{cut} (Ry) Ga		E_{cut} (Ry) In	
As	30	25		50	
P	30	25		45	
N	40	55^{nlcc}	75^{3d}	55^{nlcc}	4d: 65^{4d}

Monkhost-Pack meshes: generating **k** point ($\frac{1}{2}\frac{1}{2}\frac{1}{2}$), folding: ($4 \times 4 \times 4$)

---
[1] using atomic units

Displacements for force calculations: $\Delta d = 0.05$ Bohr. $d = (x, y, z)$

Energy convergence parameters: $\Delta E \leq 1e^{-9}$ Ha/atom.

Bibliography

[1] Sidney Yip, *Handbook of Materials Modeling* (Springer, Dordrecht, 2005).

[2] P. Hohenberg, W. Kohn, "Inhomogeneous Electron Gas", *Phys. Rev.* **136**, B864 – B871 (1964).

[3] W. Kohn, L.J. Sham, "Self-Consistent Equations Including Exchange and Correlation Effects", *Phys. Rev.* **140**, A1133 – A1138 (1965).

[4] W.L. Briggs, Van Emden Henson, S.F. McCormick, *A Multigrid Tutorial* (SIAM, 2000).

[5] T. Frauenheim, G. Seifert, M. Elsterner, Z. Hajna,l G. Jungnickel, D. Porezag, S. Suhai, R. Scholz, "A Self-Consistent Charge Density-Functional Based Tight-Binding Method for Predictive Materials Simulations in Physics, Chemistry and Biology", *phys. stat. sol.(b)* **217**, 41 – 62 (2000).

[6] M. Elstner, D. Porezag, G. Jungnickel, J. Elsner, M. Haugk, T. Frauenheim, "Self-consistent-charge density-functional tight-binding method for simulations of complex materials properties", *Phys. Rev. B* **58**, 7260 – 7268 (1998).

[7] T. Frauenheim, G. Seifert, M. Elstner, T. Niehaus, C. Köhler, M. Amkreutz, M. Sternberg, Z. Hajnal, A. Di Carlo, S. Suhai, "Atomistic simulations of complex materials: ground-state and excited-state properties", *J. Phys.: Condens. Matter* **14**, 3015 (2002).

[8] N. L. Allinger, M. A. Miller, D. H. Wertz, *J. Am. Chem. Soc. 93, 1637 (1971)* **93**, 1637 (1971).

[9] N. L. Allinger, *J. Am. Chem. Soc.* **99**, 3279 (1977).

[10] N. L. Allinger, K. Chen, J.-H. Lii, *J. Comput. Chem.* **14**, 642 (1996).

[11] J.-H. Lii, N. L. Allinger, *J. Am. Chem. Soc.* **111**, 8566 (1989).

[12] J.-H. Lii, N. L. Allinger, *J. Am. Chem. Soc.* p. 8576 (1989).

[13] N. L. Allinger, K. Chen, J.-H. Lii, *J. Comput. Chem.* **14**, 642 (1996).

[14] N. Nevins, K. Chen, N.L. Allinger, *J. Comput. Chem.* **14**, 669 (1996).

[15] N. Nevins, J.-H. Lii, N.L Allinger, *J. Comput. Chem.* **14**, 695 (1996).

[16] N. L. Allinger, K. Chen, J. A. Katzeellenbogen, S. R. Wilson and G. M. Anstead, *J. Comput. Chem.* **14**, 747 (1996).

[17] S. J. Weiner, P. A. Kollman, D. A. Case, U. C. Sing,h C. Chio, G. Alagona, S. Profeta, P. Weiner, *J. Am. Chem. Soc.* p. 106 (1984).

[18] W.D. Cornell, P. Cieplak, C.I. Bayly, I.R. Gould, K.M. Merz, D.M. Ferguson, T. Fox, J.W. Caldwell, P.A. Kollman, *J. Am. Chem. Soc.* **117**, 5179 (1995).

[19] R. Brooks, R. E. Bruccoleri, B. D. Olafson, D. J. States and S. Swaminathan, M. Karplus, *J. Comput. Chem.* **4**, 1234 (1983).

[20] Achi Brandt, "Multi-Level Adaptive Solutions to Boundary-Value Problems", *Math. Comp* **31**, 333 (1977).

[21] D. W. Brenner, "Thr Art and Science of an Analytic Potential", *phys. stat. sol. (b)* **217**, 23 (2000).

[22] O. Rioul, M. Vetterli, "Wavelets and signal processing", *IEEE Signal Processing Magazine* p. 14 (1991).

[23] B. Engquist, Z. Huang, "Heterogeneous multi-scale method: a general methodology for multi-scale modeling", *Phys. Rev. B* **67**, 092 101 (2003).

[24] M. Katsoulakis, A.J. Majda, D. G. Vlachos, "Coarse-grained stochastic processes for lattice systems", *Proc. Natl. Acad. Sci. U.S.A.* **100**, 782 (2003).

[25] G. Kresse, J. Furtmüller, *Phys. Rev. B* **54**, 11 169 (1996).

[26] G. Kresse, J. Hafner, "Ab initio molecular dynamics for liquid metals." *Phys. Rev. B* **47**, 558 – 561 (1993).

[27] G. Kresse, J. Haffner, "Ab initio molecular-dynamics simulation of the liquid-metal-amorphous-semiconductor transition in germanium", *Phys. Rev. B* **49**, 14 251 – 14 269 (1994).

[28] G. Kresse, J. Furthmüller, "Efficiency of ab-initio total energy calculations for metals and semiconductors using a plane-wave basis set." *Comp. Mat. Sci.* **6**, 15 – 50 (1996).

[29] X. Gonze, J.M. Beuken, R. Caracas, F. Detraux, M. Fuchs and G.M. Rignanese, L. Sindic, M. Verstraete, G. Zera,h F. Jollet, M. Torrent, A. Roy, M. Mikami, Ph. Ghosez and J.Y. Raty, D.C. Allan, "First-principles computation of material properties : the ABINIT software project." *Comp. Mat. Sci.* **25**, 478 – 492 (2002).

[30] M. Bockstedte, A. Kley, J. Neugebauer, M. Scheffler, *Comp. Phys. Comm.* **107**, 187 (1997).

[31] P.E. Blöchl, "Projector augmented-wave method", *Phys. Rev. B* **50**, 17 953 – 17 979 (1994).

[32] K. Schwarz, P. Blaha, "Solid state calculations using WIEN2k", *Comp. Mat. Sci.* **28**, 259 – 273 (2003).

[33] B. Delley, "From molecules to solids with the DMol3 approach", *J. Chem. Phys.* **113** (2000).

[34] M.J. Frisch, G.W. Trucks, H.B. Schlegel, G.E. Scuseria and M.A. Robb, J.R. Cheeseman, J.A. Montgomery, Jr., T. Vreven, K.N. Kudin, J.C. Burant, J.M. Millam, S.S. Iyengar and J. Tomasi, V. Barone, B. Mennucci, M. Cossi, G. Scalmani and N. Rega, G.A. Petersson, H. Nakatsuji, M. Hada, M. Ehara, K. Toyota, R. Fukuda, J. Hasegawa, M. Ishid,a T. Nakajima, Y. Honda, O. Kitao, H. Nakai, M. Klen,e X. Li, J.E. Knox, H.P. Hratchian, J.B. Cross, V. Bakken and C. Adamo, J. Jaramillo, R. Gomperts, R.E. Stratmann and O. Yazyev, A.J. Austin, R. Cammi, C. Pomelli, J.W. Ochterski, P.Y. Ayala, K. Morokuma, G.A. Voth, P. Salvador and J.J. Dannenberg, V.G. Zakrzewski, S. Dapprich, A.D. Daniels, M.C. Strain, O. Farkas, D.K. Malick, A.D. Rabuck and K. Raghavachari, J.B. Foresman, J.V. Ortiz, Q. Cu,i A.G. Baboul, S. Clifford, J. Cioslowski, B.B. Stefano,v G. Liu, A. Liashenko, P. Piskorz, I. Komaromi, R.L. Martin and D.J. Fox, T. Keith, M A. Al-Laham, C.Y. Peng, A. Nanayakkara, M.

Challacombe, P.M.W. Gill, B. Johnso,n W. Chen, M.W. Wong, C. Gonzalez, J.A. Pople, "Gaussian 03", Tech. rep., Gaussian, Inc., Wallingford CT (2004).

[35] P.A.M. Dirac, *The Principles of Quantum Mechanics* (Oxford University Press, London, 1958).

[36] L. Kleinman, D.M. Bylander, "Efficacious Form for Model Pseudopotentials", *Phys. Rev. Lett.* **48**, 1425 – 1428 (1982).

[37] B. Grabowski, T. Hickel, J. Neugebauer, "Ab initio study of the thermodynamic properties of non-magnetic elementary fcc metals: Exchange-correlation-related error bars and chemical trends", *Phys. Rev. B* **76**, 024 309 (2007).

[38] D.M. Ceperley, B.J. Alder, "Ground State of the Electron Gas by a Stochastic Method", *Phys. Rev. Lett.* **45**, 566 – 569 (1980).

[39] J.P. Perdew, K. Burke, M. Ernzerhof, *Phys. Rev. Lett.* **77**, 3865 – 3868 (1996).

[40] P. Deus, H.A. Schneider, U. Voland, K. Stiehler, "Low Temperature Thermal Expansion of InP", *phys. stat. sol. (a)* **103**, 443 (1987).

[41] T. Soma, J. Satoh, H. Matsuo, *Solid State Commun. 42* **42**, 889 (1982).

[42] P. Deus, U. Voland, H.A. Schneider, "Thermal Expansion of GaP within 20 to 300K", *phys. stat. sol. (a)* **80**, K29 (1983).

[43] K. Haruna, H. Maeta, K. Ohashit, T. Koike, "The negative thermal expansion coefficient of GaP crystal at low temperatures", *J. Phys. C* **19**, 5149 (1986).

[44] J.C. Slater, "Wave Functions in a Periodic Potential", *Phys. Rev.* **51**, 846 – 851 (1937).

[45] F. Schwabl, *Quantenmechanik 1* (Springer Verlag Berlin, Heidelberg, New York, 1988).

[46] K. Schwarz, "DFT calculations of solids with LAPW, WIEN2k", *J. Sol. Stat. Chem.* **176**, 319 – 328 (2003).

[47] N. W. Ashcroft, N. D. Mermin, *Solid State Physics* (Saunders College Publishing, Philadelphia, 1976).

[48] A. Gross, *Theoretical Surface Science. A Microscopic Perspective* (Springer, Berlin, 2003).

[49] G. Baym, *Lectures on Quantum Mechanics* (Benjamin/Cummings, Merlo Park, 1973).

[50] R. G. Parr, W. Yang, *Density-Functional Theory of Atoms and Molecules* (Oxford University Press, New York, 1989).

[51] M. Born, R. Oppenheimer, "Zur Quantentheorie der Molkeln", *Ann. Phys.* **84**, 457 (1927).

[52] W. Nolting, *Grundkurs: Theoretische Physik*, Vol. 5 (Verlag Zimmermann-Neufang, Ulmen, 1992).

[53] D. R. Hartree, *Proc. Camb. Phil. Soc.* **24**, 89 (1928).

[54] V. Fock, *Z. Phys.* **61**, 126 (1930).

[55] A. Szabo, N. S. Ostlund, *Modern Quantum Chemistry: Introduction to Advanced Electronic Structure Theory* (McGraw Hill, 1989).

[56] L.H. Thomas, "The calculation of atomic fields", *Proc. Camb. Phil. Soc.* **23**, 542 (1927).

[57] E. Fermi, "Un metodo statistica per la determinazione di alcune priorieta dell'atomie", *Atti Della Reale Accademia Nazionale Dei Lincei* **6**, 602 (1927).

[58] E. Fermi, "Eine statistische Methode zur Bestimmung iniger Eigenschaften des Atoms und ihre Anwendung auf die Theorie des periodischen Systems der Elemente", *Z. Phys.* **48**, 73 (1928).

[59] R. M. Dreizler, E. K. U. Gross, *Density Functional Theory* (Springer, Berlin, 1990).

[60] A. Dick, *An-initio STM and STS Simulations on Magnetic and Nonmagnetic Metallic Surfaces*, Ph.D. thesis, University of Paderborn (2008).

[61] M. C. Payne, M. P. Teter, D. C. Allen, T. A. Aria,s J. D. Joannopoulos, "Iterative minimization techniques for ab-initio total-energy calculations: molecular dynamics and conjugate-gradients", *Rev. Mod. Phys.* **64**, 1045 – 1097 (1992).

[62] H. J. Monkhorst, J. D. Pack, "Special points for Brillouin-zone integrations", *Phys. Rev. B* **13**, 5188 – 5192 (1976).

[63] D. J. Chadi, M. L. Cohen, *Phys. Rev. B* **8**, 5747 (1973).

[64] A. Baldereschi, *Phys. Rev. B* **7**, 5212 (1973).

[65] G. Lehmann, M. Taut, *Phys. Stat. Sol. B* **54**, 469 (1972).

[66] M. Methfessel, A. T. Paxton, "High-precision sampling for Brillouin-zone integration in metals", *Phys. Rev. B* **40**, 3616 – 3621 (1989).

[67] Arias, "New Algebraic Formulation of Density Functional Calculation", *Comp. Phys. Comm.* **128**, 1 (2000).

[68] P.P. Ewald, *Ann. Phys.* **54**, 519 (1917).

[69] P.P. Ewald, *Ann. Phys.* **54**, 557 (1917).

[70] P.P. Ewald, *Ann. Phys.* **64**, 253 (1921).

[71] I.N. Bronstein, K.A. Semedjajew, G. Musiol H. Muühlig, *Taschenbuch der Mathematik* (Verlag Harri Deutsch, 1993).

[72] U. von Barth, C.D. Gelatt, "Validity of the frozen-core approximation and pseudopotential theory for cohesive energy calculations", *Phys. Rev. B* **21**, 2222 – 2228 (1980).

[73] X. Gonze, R. Stumpf, M. Scheffler, "Analysis of separable potentials", *Phys. Rev. B* **44**, 8503 – 8513 (1991).

[74] C.G. van de Walle, P.E. Blöchl, "First-principles calculations of hyperfine parameters", *Phys. Rev. B* **47**, 4244 – 4255 (1993).

[75] O.K. Andersen, *Phys. Rev. B* **12**, 3060 (1975).

[76] D.J. Singh, *Plane waves, pseudopotentials and the LAPW method* (Kluwer Academic Publisher, Bosten, Dortrecht, London, 1994).

[77] H.L. Skriver, *The LMTO Method* (Springer-Verlag, 1984).

[78] H.J.F. Jansen and A.J. Freeman, "Total-energy full-potential linearized augmented-plane-wave method for bulk solids: Electronic and structural properties of tungsten", *Phys. Rev. B* **30**, 561 – 569 (1984).

[79] D. Singh, "Ground-state properties of lanthanum: Treatment of extended-core states", *Phys. Rev. B* **43**, 6388 – 6392 (1991).

[80] P.E. Blöchl, C.J. Först, J. Schimpl, "The Projector augmented wave method: ab-initio molecular dynamics with full wavefunctions", *Bull. Mater. Sci.* **26**, 33 – 50 (2003).

[81] P.E. Blöchl, J. Kästner, C.J. Först, *Electronic structure methods: Augmented Waves, Pseudopotentials and the Projector Augmented Wave method*, Vol. 1 (Springer-Verlag, 2005).

[82] J. C. Slater, G. F. Koster, *Phys. Rev.* **94**, 1498 (1954).

[83] R. Barret, M. Berry, T.F. Chan, J. Demmel, J. Donato and J. Dongarra, V. Eijkhout, R. Pozo, C. Romine, H. van der Vorst, *Templates for the Solution of Linear Systems: Building Blocks for Iterative Methods* (SIAM, 1993).

[84] Z. Bai, J. Demmel, J. Dongarra, *Templates for the Solution of Algebraic Eigenvalue Problems: A Practical Guide* (SIAM, 2000).

[85] J. K. Cullum, R. A. Willoughby, "Computing eigenvalues of very large symmetric matrices-an implementation of a Lanczos algorithm with no reorthogonalization", *J. Comp. Phys.* **44**, 329 (1981).

[86] J. K. Cullum, R. A. Willoughby, *Lanczos algorithms for Large Symmetric Eigenvalue Computations. Volume 1, Theory* (Birkhäuser, Boston, 1985).

[87] W.M.C. Foulkes, R. Haydock, "Tight-binding models and density-functional theory", *Phys. Rev. B* **39**, 12 520 – 12 536 (1989).

[88] R. Feynman, "Forces in Molecules", *Phys. Rev.* **56**, 340 (1937).

[89] A.C. Hurley, "The electrostatic calculation of molecular energies. 1. Methods of calculating molecular energies", *Proc. R. Soc. London, Ser. A* **226**, 170 – 178 (1954).

[90] M. Di Ventra, S.T. Pantelides, "Hellmann-Feynman theorem and the definition of forces in quantum time-dependent and transport problems", *Phys. Rev. B* **61**, 16 207 – 16 212 (2000).

[91] Ch. Kittel, *Introduction to Solid State Physics* (Wiley, 1986), 6th edition edn.

[92] W. H. Press, S. A. Teulosky, W. T. Vetterling, B. P. Flannery, *Numerical Recipes in C: The art of scientific computing* (Cambridge University Press, 1992), 2nd ed. edn.

[93] A. Williams, J. Soler, *Bull. Am. Phys. Soc.* **32**, 562 (1987).

[94] M.R. Hestenes, E. Stiefel, "Methods of conjugate gradients for solving linear systems", *J. Research Nat. Bur. Standard* **49**, 409 – 436 (1952).

[95] G.W. Pratt, "Wave Functions and Energy Levels for Cu+ as Found by the Slater Approximation to the Hartree-Fock Equations", *Phys. Rev.* **88**, 1217 – 1224 (1952).

[96] P. Pulay, "Convergence acceleration of iterative sequences. the case of scf iteration", *Chem. Phys. Lett.* **73**, 393 – 398 (1980).

[97] G. P. Kerker, "Efficient iteration scheme for self-consistent pseudopotential calculations", *Phys. Rev. B* **23**, 3082 (1981).

[98] D. Raczkowski, A. Canning, L.W. Wang, "Thomas-Fermi charge mixing for obtaining self-consistency in density functional calculations", *Phys. Rev. B* **64**, 121 101 (2001).

[99] R. Fletcher, *Practical Methods of Optimization* (Wiley, 1981).

[100] P.v.Rague Schleyer, *Encyclopedia of Computational Chemistry* (John Wiley and Sons, 1998).

[101] L. Verlet, "Computer "Experiments" on Classical Fluids. I. Thermodynamical Properties of Lennard-Jones Molecules", *Phys. Rev.* **159**, 98 – 103 (1967).

[102] L. Verlet, "Computer "Experiments" on Classical Fluids. II. Equilibrium Correlation Functions", *Phys. Rev.* **165**, 201 – 214 (1968).

[103] S. Nose, "A molecular dynamics method for simulations in the canonical ensemble", *Molec. Phys.* **52**, 255 – 268 (1984).

[104] Mermin, *Phys. Rev.* **137**, A1441 (1965).

[105] D. C. Wallace, *Thermodynamics of Crystals* (Dover Publications, Inc., Mineolta, New York, 1998).

[106] S. Baroni, P. Giannozzi, A. Testa, *Phys. Rev. Lett.* **58**, 1861 (1987).

[107] R. Resta, *Festkö rperprobleme: Advances in Solid State Physics*, Vol. 25 (Vieweg, Braunschweig, 1985).

[108] P. Giannozzi, S. de Gironcoli, P. Pavone, S. Baroni, *Phys. Rev. B* **43**, 7231 (1991).

[109] R. Heid, K.P. Bohnen, K.M. Ho, "Ab initio phonon dynamics of rhodium from a generalized supercell approach", *Phys. Rev. B* **59**, 7407 (1998).

[110] P.Y.Yu, M.Cardona, *Fundamentals of Semiconductors: Physcis and Materials Properties* (Spinger Verlag, Berlin, 1996), p. 104.

[111] M. Born, W. Heisenberg, P. Jordan, "Zur Quantenmechanik II", *Z. Phys.* **35**, 557 – 615 (1925).

[112] U. Breymann, "Geprüfte Dimensionen", *iX* **11**, 174 – 180 (1994).

[113] http://www.boost.org .

[114] X. Gonze, *Phys. Rev. A* **52**, 1086 (1995).

[115] M. Städele, J.A. Majewski, P. Vogl, A. Görling, "Exact Kohn-Sham Exchange Potential in Semiconductors", *Phys. Rev. Lett.* **79**, 2089 – 2092 (1997).

[116] M.C. Payne, J.D. Joannopoulos, D.C. Allan, M.P. Tete,r D.H. Vanderbuilt, "Molecular Dynamics and ab initio Total Energy Calculations", *Phys. Rev. Lett.* **56**, 2656 – 2656 (1986).

[117] S.V. Novikov, N.M. Stanton, R.P. Campion, R.D. Morris, H.L. Geen, C.T. Foxon, A.J. Kent, "Growth and characterization of free-standing zinc-blende (cubic) GaN layers and substrates", *Semicond. Sci. Technol.* **23**, 015 018 (2008).

[118] C. Stampfl, C. G. Van de Walle, *Phys. Rev. B* **59**, 5521 (1998).

[119] M. Fuchs, J.L.F. da Silva, C. Stampfl, J. Neugebauer, M. Scheffler, "Cohesive properties of group-III nitrides: A comparative study of all-electron and pseudopotential calculations using the generalized gradient approximation", Phys. Rev. B **65**, 245 212 (2002).

[120] M. Fuchs, M. Scheffler, Comp. Phys. Comm. **119**, 67 (1999).

[121] M. Alouani, J.M. Wills, Phys. Rev. B **54**, 2480 (1996).

[122] M. Leszczynski, V.B. Pluzhnikov, A. Czopnik, J. Bak-Misiuk and T. Slupinski, J. Appl. Phys. **82**, 4678 (1997).

[123] R.I. Cottam, G.A. Saunders, J. Phys. C **6**, 2105 (1973).

[124] S.T. Weir, Y.K. Vohra, C.A. Vanderborgh, A.L. Ruoff, Phys. Rev. B **39**, 1280 (1989).

[125] H. Arabi, A. Pourghazi, F. Ahmadian, Z. Nourbakhsh, "First-principles study of structural and electronic properties of different phases of GaAs", Physica B **373**, 16 (2006).

[126] Landolt-Börnstein, Condensed Matter, III/41A1a (2001).

[127] S. Biernacki, M. Scheffler, "Negative Thermal Expansion of Diamond and Zinc-Blende Semiconductors", Phys. Rev. Lett. **3**, 291 (1989).

[128] B.K. Tanner, A.G. Turnbull, C.R. Stanley, A.H. Kea,n M. McElhinney, Appl. Phys. Lett. 59 **59**, 2272 (1991).

[129] J.V. Ozolin'sh, G.K. Averkieva, A.F. Ilvin'sh, N.A. Goryunova, Sov. Phys. Cryst. (English Transl.) **7**, 691 (1963).

[130] R.G. Greene, H. Luo, A.L. Ruoff, J. Appl. Phys. **76**, 7296 (1994).

[131] Y.K. Vohra, S.T. Weir, A.L. Ruoff, Phys. Rev. B **31**, 7344 (1985).

[132] V.N. Bessolov, S.G. Konnikov, V.I. Umanskii, Yu.P. Yakovlev, Sov. Phys. Solid State (English Transl.) 24 (1982) 875 **24**, 875 (1982).

[133] C.O. Rodriguez, R.A. Casali, E.I. Peltzer, O.M. Cappannini and M. Methfessel, Phys. Rev. B **40**, 3975 (1992).

[134] A. Polian, J.P. Itie, C. Jaubertie-Carillon, A. Dartyge and A. Fontaine, H. Tolentino, High Pressure Res. **4**, 309 (1990).

[135] C.S. Menoni, I.L. Spain, Phys. Rev. B **35**, 7520 (1987).

[136] J.P. Itie, A. Polian, C. Jauberthie-Carillon, E. Dartyge and A. Fontaine, H. Tolentino, G. Tourillon, Phys. Rev. B **40**, 9709 (1989).

[137] R.W.G. Wyckoff, Semiconductors: Data Handbook, 2nd. Edition, Krieger 1986 (Springer, Berlin, 2004), chap. Crystal Structures.

[138] R. Ahmed, H. Akbarzadeh, Fazal-e-Aleem, "Ab-initio Study of Structural Properties of III- Nitrides", in MODERN TRENDS IN PHYSICS RESEARCH: Second International Conference on Modern Trends in Physics Research MTPR-06 **888**, 42 (2007).

[139] M.E. Sherwin, T.J. Drummond, "J. Appl. Phys. 69, 8423 (1991)." J. Appl. Phys. **69**, 8423 (1991).

[140] A. Garcia, C. Elsässer, J. Zhu, S.G. Louie, M.L. Cohen, *Phys. Rev. B* **46**, 9829 (1992).

[141] A. Garcia, C. Elsässer, J. Zhu, S.G. Louie, M.L. Cohen, *Phys. Rev. B* **47**, 4150(E) (1993).

[142] M. Buongiorno Nardelli, K. Rapcewicz, E. L. Briggs, C. Bun- garo, J.Bernholc, "III-V Nitrides", in *MRS Symposia Proceedings*, F. A. Ponce, T. D. Moustakas, I. Akasaki„ B. A. Monemar, ed., **449**, 893 (1997).

[143] A. Onton, "10th Int. Conf. on Physics of Semiconductors", p. 107 (1970).

[144] Monemar, *J. Appl. Phys.* **50**, 4362 (1979).

[145] D. Strauch, H. Dorner, *J. Condens. Matter* **2**, 1457 (1990).

[146] N.S. Orlova, *Phys. Status Solidi (b)* **119**, 541 (1983).

[147] R. Carles, N. Saint-Cricq, J.B. Renucci, M.A. Renucc,i A. Zwick, *Phys. Rev. B* **22**, 4804 (1980).

[148] Ch. Eckl, P. Pavone, J. Fritsch, U. Schröder, *The Physics of Semiconductors* (Singapore: World Scientific, 1996), Vol. 1, p. 229.

[149] T. Pletl, P. Pavone, Ul.E. Dieter Strauch, "First-principles study of lattice-dynamical and elastic trends in tetrahedral semiconductors", *Physica B* **263-264**, 392–395 (1999).

[150] J.L. Yarnell, J.L. Warren, R.G. Wenzel, P.J. Dean, *Neutron Inelastic Scattering* (International Atomic Energy Agency, Vienna, 1968), p. 301.

[151] P.H. Borcherds, K. Kunc, G.F. Alfrey, R.L. Hall, *J. Phys. C Solid State Phys.* **12**, 4699 (1979).

[152] B. Podor, "Zone Edge Phonons in Gallium Phosphide", *phys. stat. sol. (b)* **120**, 207 (1983).

[153] P.H. Borcherds, G.F. Alfrey, D.H. Saunderson, A.D.B. Woods, *J. Phys. C* **8**, 2022 (1975).

[154] A. Mooradian, G.B. Wright, *Solid State Commun.* **4**, 431 (1966).

[155] J. Fritsch, P. Pavone, U. Schröder, *Phys. Rev. B* **52**, 11 326 (1995).

[156] K. Karch, J.M. Wagner, F. Bechstedt, *Phys. Rev. B* **57**, 7043 (1998).

[157] F. Bechstedt, U. Grossner, J. Fruthmueller, *Phys. Rev. B* **62**, 8003 (2000).

[158] A. Cros, R. Dimitrov, H. Ambacher, M. Stutzmann, S.Christiansen and M. Albrecht, H.P. Strunk, *J. Crystal Growth* **181**, 197 (1997).

[159] Rashid Ahmed, H. Akbarzadeh, Fazal-e-Aleemaam, "A first principle study of band structure of III-nitrides compounds", *Physica B* **370**, 52 (2005).

[160] *Properties of Group IV, III-V and II-VI Semiconductors* (Wiley, England, 2005).

[161] A. L. da Rosa, *Density Functional Theory Calculations on Anti-Surfactants at GaN Surfaces*, Ph.D. thesis, TU Berlin (2003).

[162] R. Miotto, G. P. Srivastava, A. C. Ferraz, *Phys. Rev. B* **59**, 3008 (1999).

[163] A. Tabata, A.P. Lima, L.K. Teles, L.M. Scolfaro, R.R. Leite, V. Lemos, B. Schöttker, T. Frey, D. Schikora and K. Lischka, *Appl. Phys. Lett.* **74**, 362 (1999).

[164] H.M.T. Tütüncü, G.P. Srivastava, S. Dumana, "Lattice dynamics of the zinc-blende and wurtzite phases of nitrides", *Physica B* **316-317**, 190–194 (2002).

[165] P. Rinke, M. Winkelnkemper, A. Qteish, D. Bimber, J. Neugebauer, M. Scheffler, "Consistent set of band parameters for the group-III nitrides AlN, GaN, and InN", *Phys. Rev. B* **77**, 075202 (2008).

[166] S. Adachi, *Properties of Group-IV, III-V and II-VI Semiconductors* (Willey, England, 2005).

[167] K. Karch, F. Bechstedt, T. Pletl, "Lattice dynamics of GaN: Effects of 3d electrons", *Phys. Rev. B* **56**, 3560 (1997).

[168] H.M. Kagaya, T. Soma, *Solid State Commun.* **62**, 707 (1987).

[169] H.M. Kagaya, T. Soma, *Phys. Status Solidi (b)* **142**, 411 (1987).

[170] A. Debernardi, *Solid State Commun.* **113**, 1 (2000).

[171] T.F. Smith, G.K. White, *J. Phys. C* **5**, 2031 (1975).

[172] P.W. Sparks, C.A. Swenson, *Phys. Rev.* **163**, 779 (1967).

[173] S.N. Novikova, *Sov. Phys. Solid State (Engl. Trans.)* **3**, 129 (1961).

[174] N.N. Sirota, L.I. Berger, *Inzh. Fiz. Zhurnal, Akad. Nauk. Beloruss. SSR* **2**, 104 (1959).

[175] H.G. Grimmeiss, B. Monemar, *Phys. Status. Solidi. (a)* **5**, 109 (1971).

[176] H. Maeta, T. Kat, S. Okuda, *J. Appl. Crystallogr.* **9**, 378 (1976).

[177] V.M. Glazov, A.S. Pashinkin, "Thermal expansion and Heat Capacity of GaAs and InAs", *Inorg. Materials* **36**, 289 (2000).

[178] Lu Lai-Yu, Chen Xiang-Rong, Yu Bai-Ru, Gou Qing-Quan, "First-principles calculations for transition phase and thermodynamic properties of GaAs", *Chinese Physics* **15**, 802 (2006).

[179] V.P. Vasil'ev, J.-E. Gachon, "Thermodynamic Properties of InP", *Inorg. Materials* **42**, 1287 (2006).

[180] I. Ansara, C. Chatillon, H.L.Lukas, "A Binary Data Base for III-V Compounds Semiconductor Systems", *CALPHAD: Comp. Cpupling Phase Diagrams Thermochem.* **18**, 177 (1994).

[181] J. Jacobs, "Phosphorus at High Temperature and Pressure", *J. Chem. Phys.* **5**, 945 (1937).

[182] W.S. Holmes, "Heat of Combustion of Phosphorus and the Enthalpies of Formation of P4O10 and H3PO4", *Trans. Faraday Soc.* **58**, 916 (1962).

[183] P.A.G. O'Hare, W.N. Hubbard, "Fluorine Bomb Calometry", *Trans. Faraday Soc.* **62**, 2709 (1966).

[184] P.A.G. O'Hare, B.M. Levis, "Thermodynamic Stability of Orthorhombic Black Phosphoruus", *Thermochim. Acta* **129**, 57 (1988).

[185] I. Yamaguchi, K. Itogaki, A. Iazawa, "Measurements Heat of Formation of GaP, InP, GaAs, InAs, and InSb", *mater. Trans. JIM* **35**, 596 (1994).

[186] L. Ismer, *Protonentransport in Wasserstoffbrückenbindungen*, Master's thesis, Technische Universität Berlin (2002).

[187] L. Ismer, J. Ireta, S. Boeck, J. Neugebauer, "Phonon spectra and thermodynamic properties of the in polyalanine alpha helix: A density-functional-theory-based harmonic vibrational analysis", *Phys. Rev. E* **71**, 031911-1 (2005).

[188] L. Ismer, J. Ireta, J. Neugebauer, "First principles free energy analysis of helix stability: The origin of the low pi-helices", *J. Phys. Chem. B* **112**, 4109 (2008).

[189] L. Ismer, *First principles based thermodynamic stability analysis of the secondary structure of proteins*, Ph.D. thesis, University of Paderborn (2009).

[190] M. Petrov, L. Lymperakis, M. Friak, J. Neugebauer, "Ab-initio based conformational study of the crystalline alpha-chitin", *to be submitted* .

[191] B. Grabowski, L. Ismer, T. Hickel, J. Neugebauer, "Ab initio up to the melting point: Anharmonicity and vacancies in aluminum", *Phys. Rev.* **79**, 134106 (2009).

[192] Blazej Grabowski, *Ab-initio based free-energy surfaces: Method development and application to aluminum an iron*, Master's thesis, University of Paderborn (2005).

[193] O. Marquardt, D. Mourad, S. Schulz, T. Hickel, G. Czycholl and J. Neugebauer, "A comparison of atomistic and continuum theoretical approaches to determine electronic properties of GaN/AlN quantum dots", *Phys. Rev. B* **78**, 235302 (2008).

[194] T. Hammerschmidt, P. Kratzer, M. Scheffler, "Analytic many-body potential for InAs/GaAs surfaces and nanostructures: Formation energy of InAs quantum dots", *Phys. Rev. B* **77**, 235303 (2008).

[195] T. D. Young, O. Marquardt, *phys. stat. sol. (c)* **6**, 557 (2009).

[196] M. Albrecht, L. Lymperakis, J. Neugebauer, J.E. Northrup, L. Kirste, M. Leoux, I. Grzegory, S. Porowski, "Chemically ordered AlxGa1-xN alloys: Spontaneous formation of natural quantum wells", *Phys. Rev. B* **71**, 035314 (2005).

[197] O. Marquardt, T. Hickel, J. Neugebauer, C.G. van de Walle, "Influence of polarization effects due to thickness fluctuations in nonpolar InGaN/GaN quantum wells", *to be published* .

[198] O. Marquardt, T. Hickel, J. Neugebauer, "Polarization-induced charge carrier separation in polar and nonpolar grown GaN quantum dots", *to be published* .

[199] J. Kioseoglou, E. Kalesaki, Ph. Komninou, Th. Karakostas and L. Lymperakis, J. Neugebauer, "Electronic structure of 1/6<2023> partial dislocations in wurtzite GaN." *to be submitted* .

[200] H. Abu-Farsakh, *Maximally-localized Wannier functions in III-V Semiconductors*, Master's thesis, Yarmouk University, Irbid, Jordan (2003).

[201] T. Hammerschmidt, M. A. Migliorato, D. Powell, A. G. Cullis and G. P. Srivastava, "Composition and Strain Dependence of the Piezoelectric Coefficients in Semiconductor Alloys", in *MRS Proceedings* (2007).

[202] H. Abu-Farsakh, A. Qteish, "Ionicity scale based on the centers of maximally localized Wannier functions", *Phys. Rev. B* **75**, 085201 (2007).

[203] H. Abu-Farsakh, J. Neugebauer, "Enhancing nitrogen solubility in GaAs and InAs by surface kinetics: An ab initio study", *Phys. Rev. B* **79**, 155311 (2009).

[204] C. Freysoldt, J. Neugebauer, C. van de Walle, "Fully ab initio finite-size corrections for charged defect supercell calculations", *Phys. Rev. Lett.* **102**, 035 702 (2009).

[205] S. Frank, *Einfluss der Materialeigenschaften auf den Ferromagnetismus von GaMnAs*, Master's thesis, University Ulm (2006).

[206] P. Rinke, A. Qteish, J. Neugebauer, C. Freysoldt, M. Scheffler, "Structural phase transformation of GaN under high-pressure: an exact exchange study", *New J. Phys.* **7**, 2126 (2005).

[207] P. Rinke, M. Scheffler, A. Qteish, M. Winkelkemper, D. Bimberg, "Band gap and band parameters of InN and GaN from quasiparticle energy calculations based on exact-exchange density-functional theory", *Appl. Phys. Lett.* **89**, 161 919 (2006).

[208] "Combining GW calculations with exact-exchange density-functional theory: an analysis of valence-band photoemission for compound seminconductors", *New J. Phys.* **7**, 126 (2005).

[209] A. Qteish, A.I. Al-Sharif, M. Fuchs, M. Scheffler, S. Boeck, J. Neugebauer, "Exact-exchange calculations of the electronic structure of AlN, GaN and InN", *Comp. Phys. Comm.* **169**, 28 (2005).

[210] Abdallah Qteish, Patrick Rinke, Matthias Scheffler, Joerg Neugebauer, "Exact-exchange based quasi-particle energy calculations for the band gap, effective masses and deformation potentials of ScN", *Phys. Rev. B* **74**, 245 208–1 (2006).

[211] A. Qteish, A.I. Al-Sharif, M. Fuchs, M. Scheffler, S. Boeck, J. Neugebauer, "Role of semicore states in the electronic structure of group-III nitrides: An exact exchange study", *Phys. Rev. B* **72**, 155 317 (2005).

[212] A.I. Al-Sharif, "Structural phase transformation of GaN under high-pressure: an exact exchange study", *Sol. Stat. Comm.* **135**, 515 (2005).

[213] M. Wahn, J. Neugebauer, "Generalized Wannier functions: An efficient way to construct ab-initio tight-binding parameters for group-III nitrides", *phys. stat. solidi (b)* **243**, 1583 (2006).

[214] C. Freysoldt, S. Boeck, J. Neugebauer, "Direct minimization technique for metals in density-functional theory", *Phys. Rev. B* .

[215] O. Marquardt, S. Boeck, C. Freysoldt, T. Hickel, J. Neugebauer, "Implementation of the real-space k.p formalism and continuum elasticity theory in the plane-wave software library S/PHI/nX", *submitted to Comp. Phys. Comm.* (2009).

[216] M. Todorova, L. Ismer, J. Neugebauer, "Role of anharmonic contributions for the elasticity of ice", *in prep.* .

[217] A. Smith, R. Yang, H.Q. Yang, W.R.L. Lambrecht, A. Dick and J. Neugebauer, "Aspects of spin-polarized scanning tunneling microscopy at the atomic scale: experiment, theory, and simulation", *Surface Science* **561**, 154 (2004).

[218] A.R. Smith, R. Yang, H.Q. Yang, A. Dick, J. Neugebauer and W.R.L. Lambrecht, "Recent advances in atomic-scale spin-polarized scanning microscopy", *Microscopy Research and Technology* **66**, 72 (2005).

[219] V. Bubnik, *Visualization, Modeling of Molecules, Crystals*, Master's thesis, Brno University of Technology (2008).

[220] T. Uchdorf, *Developing a general purpose database application for multiphysics*, Master's thesis, fachhochschule Aachen (2008).

[221] D. Calistea, Y. Pouillona, M.J. Verstraetea, V. Olevanoa and X. Gonze, "Sharing electronic structure and crystallographic data with ETSF_IO", *Comp. Phys. Comm.* **179**, 748 (2008).

Die VDM Verlagsservicegesellschaft sucht für wissenschaftliche Verlage abgeschlossene und herausragende

Dissertationen, Habilitationen, Diplomarbeiten, Master Theses, Magisterarbeiten usw.

für die kostenlose Publikation als Fachbuch.

Sie verfügen über eine Arbeit, die hohen inhaltlichen und formalen Ansprüchen genügt, und haben Interesse an einer honorarvergüteten Publikation?

Dann senden Sie bitte erste Informationen über sich und Ihre Arbeit per Email an *info@vdm-vsg.de*.

Sie erhalten kurzfristig unser Feedback!

VDM Verlagsservicegesellschaft mbH
Dudweiler Landstr. 99   Telefon +49 681 3720 174
D - 66123 Saarbrücken   Fax   +49 681 3720 1749
www.vdm-vsg.de

Die VDM Verlagsservicegesellschaft mbH vertritt

Printed by Books on Demand GmbH, Norderstedt / Germany